W9-CEQ-463

Chicago Guides
to *Writing*, Editing,
and Publishing

01234

THE CHICAGO GUIDE
TO WRITING ABOUT
Numbers

SECOND EDITION

JANE E. MILLER

THE UNIVERSITY OF CHICAGO PRESS

Chicago and London

JANE E. MILLER is a research professor at the Institute for Health, Health Care Policy and Aging Research and professor in the Edward J. Bloustein School of Planning and Public Policy at Rutgers University, as well as the faculty director of Project L/EARN. She is the author of *The Chicago Guide to Writing about Multivariate Analysis, Second Edition*, also from the University of Chicago Press.

The University of Chicago Press, Chicago 60637
The University of Chicago Press, Ltd., London
© 2004, 2015 by The University of Chicago
All rights reserved. First edition, 2004.
Second edition, 2015.
Printed in the United States of America

24 23 22 21 20 19 18 17 16   3 4 5

ISBN-13: 978-0-226-18563-7 (cloth)
ISBN-13: 978-0-226-18577-4 (paper)
ISBN-13: 978-0-226-18580-4 (e-book)
DOI: 10.7208/chicago/9780226185804.001.0001

Library of Congress Cataloging-in-Publication Data

The Chicago guide to writing about numbers. — Second edition / Jane E. Miller
    pages cm
   Includes bibliographical references and index.
    ISBN 978-0-226-18563-7 (cloth : alk. paper) — ISBN 978-0-226-18577-4
(pbk. : alk. paper) — ISBN 978-0-226-18580-4 (e-book)   1. Technical writing.
I. Miller, Jane E. (Jane Elizabeth), 1959–   II. Series: Chicago guides to writing,
editing, and publishing.
   T11.M485 2015

                                             2014032722

⊗ This paper meets the requirements of
ANSI/NISO Z39.48-1992 (Permanence of Paper).

To my parents,

for nurturing my

love of numbers

CONTENTS

TABLES

FIGURES

BOXES

ACKNOWLEDGMENTS

This book is the product of my experience as a student, practitioner, and teacher of quantitative analysis and presentation. Thinking back on how I learned to write about numbers, I realized that I acquired most of the ideas from patient thesis advisors and collaborators who wrote comments in the margins of my work to help me refine my presentation of quantitative information. This book was born out of my desire to share the principles and tools for writing about numbers with those who don't have access to that level of individualized attention.

Foremost, I would like to thank my mentors from the University of Pennsylvania, who planted the seeds for this book. Samuel Preston was the source of several ideas in this book and the inspiration for others. He, Jane Menken, and Herbert Smith not only served as models of high standards for communicating quantitative material to varying audiences, but taught me the skills and concepts needed to meet those standards.

Many colleagues and friends offered tidbits from their own experience that found their way into this book, or provided thoughtful feedback on early drafts. In particular, I would like to thank Deborah Carr, Diane (Deedee) Davis, Don Hoover, Ellen Idler, Tamarie Macon, Julie McLaughlin, Dawne Mouzon, Louise Russell, Usha Sambamoorthi, Tami Videon, and Lynn Warner. Susan Darley and Ian Miller taught me a great deal about effective analogies and metaphors. Jane Wilson gave invaluable advice about organization, writing, and design, while Tamara Swedberg and Jim Walden provided key support on graphical design. Kathleen Pottick, Keith Wailoo, and Allan Horwitz provided indispensable guidance and support for bringing the first edition of this book to fruition. As the director of the Institute for Health, Health Care Policy, and Aging Research at Rutgers University, David Mechanic generously granted me the time to work on this venture. Finally, I would like to thank my students for providing a steady stream of ideas about what to include in the book, as well as opportunities to test and refine the materials.

|23456789I0III2I3

INTRODUCTION

Writing about numbers is an essential skill, an important tool in the repertoire of expository writers in many disciplines. For a quantitative analysis, presenting numbers and patterns is a critical element of the work. Even for works that are not inherently quantitative, one or two numeric facts can help convey the importance or context of your topic. An issue brief about education policy might include a statistic about the prevalence of school voucher programs and how that figure has changed since the policy was enacted. Or, information could be provided about the impact of vouchers on students' test scores or parents' participation in schools. For both qualitative and quantitative works, communicating numeric concepts is an important part of telling the broader story.

As you write, you will incorporate numbers in several different ways: a few carefully chosen facts in a short article or a non-quantitative piece, a table in the analytic section of a scientific report, a chart of trends in the slides for a speech, a case example in a policy statement or marketing report. In each of these contexts, the numbers support other aspects of the written work. They are not taken in isolation, as in a simple arithmetic problem. Rather, they are applied to some larger objective, as in a math "word problem" where the results of the calculations are used to answer some real-world question. Instead of merely calculating average out-of-pocket costs of prescription medications, for instance, the results of that calculation would be included in an article or policy statement about insurance coverage for prescription medications. Used in that way, the numbers generate interest in the topic or provide evidence for a debate on the issue.

In many ways, writing about numbers is similar to other kinds of expository writing: it should be clear, concise, and written in a logical order. It should start by stating an idea or proposition, then provide evidence to support that thesis. It should include examples that the expected audience can relate to, and descriptive language that enhances their understanding of how the evidence relates to the question. It should be written at a level of detail that is consistent with its expected use. It should set the context and define terms the audience might not be expected to know, but do so in ways that distract as little as possible from the main thrust of the work. In short, it will follow many of the principles of good writing, but with the addition of quantitative information.

When I refer to writing about numbers, I mean "writing" in a broad sense: preparation of materials for oral or visual presentation as well as materials to be read. Most of the principles outlined in this book apply equally to speech writing and the accompanying slides, or to development of a Web site, a research poster, educational podcast, or automated slide show.

Writing effectively about numbers also involves *reading* effectively about numbers. To select and explain pertinent quantitative information for your work, you must understand what those numbers mean and how they were measured or calculated. The first few chapters provide guidance on important features such as units and context to watch for as you garner numeric facts from other sources.

WHO WRITES ABOUT NUMBERS

Numbers are used everywhere. In daily life, you encounter numbers in stock market reports, recipes, sports telecasts, the weather report, and many other places. Pick up a copy of your local newspaper, turn on the television, or connect to the Internet and you are bombarded by numbers being used to persuade you of one viewpoint or another. In professional settings, quantitative information is used in laboratory reports, research papers, books, and grant proposals in the physical and social sciences, policy briefs, and marketing and finance materials. Consultants and applied scientists need to communicate with their clients as well as with highly trained peers. In all of these situations, for numbers to accomplish their purpose, writers must succinctly and clearly convey quantitative ideas. Whether you are a college student or a research scientist, a policy analyst or an engineer, a journalist or a consultant, chances are you need to write about numbers.

Facility in writing about numbers is a critical element of quantitative literacy—the ability to apply mathematical reasoning and computations to address substantive issues on a wide range of topics. Books such as *Mathematics and Democracy: The Case for Quantitative Literacy* (Steen 2001) and *Achieving Quantitative Literacy: An Urgent Challenge for Higher Education* (Steen 2004) make a compelling case for the importance of quantitative literacy not only in professions such as those listed above, but also in tasks of daily life related to personal finance, citizenship, health, and other activities that require using numeric information to make decisions. However, a series of books, including seminal works by Paulos (2001), Dewdney (1996), and Best (2001) suggest that many people emerge from school ill equipped to apply quantitative literacy skills to the kinds of questions central to functioning in modern society.

Despite the apparently widespread need, few people are formally trained to write about numbers. Communications specialists learn to write for varied audiences, but rarely are taught specifically to deal with numbers. Scientists and others who routinely work with numbers learn to calculate and interpret the findings, but rarely are taught to describe them in ways that are comprehensible to audiences with different levels of quantitative expertise or interest. Moreover, although the variety of topics named above demonstrates that substantive disciplines including the social sciences, biological sciences, and history all have roles to play in developing and practicing quantitative literacy (Miller 2010), many students spend little time learning to work with numbers in such courses.

I have seen poor communication of numeric information at all levels of training and experience, from papers by undergraduates who were shocked at the very concept of putting numbers in sentences, to presentations by business consultants, policy analysts, and scientists, to publications by experienced researchers in elite, peer-reviewed journals. This book is intended to bridge the gap between correct quantitative analysis and good expository writing, taking into account the intended objective and audience.

TAILORING YOUR WRITING TO ITS PURPOSE

A critical first step in any writing process is to identify the audience and objectives of the written work, which together determine many aspects of how you will write about numbers.

Objectives

First, determine the objectives of the piece. Are you aiming to communicate a simple point in a public service announcement? To use statistics to persuade magazine readers of a particular perspective? To serve as a reference for those who need a regular source of data for comparison and calculation? To test hypotheses using the results of a complex statistical analysis?

Audience

Next, identify the audience(s) for your work, what they need to know about your results, and their level of training and comfort with numeric information. Will your readers be an eighth-grade civics class? A group of legislators who need to be briefed on an issue? A panel of scientific experts?

If you are writing for several audiences, expect to write several versions. For example, unless your next-door neighbor has a degree in statistics, chances are he will not want to see the statistical calculations you used to analyze which schools satisfy the Common Core State Standards. He might, however, want to know what your results mean for your school district—in straightforward language, without Greek symbols, standard errors, or jargon. On the other hand, if the National Science Foundation funded your research, they will want a report with all the gory statistical details and your recommendations about research extensions as well as illustrative case examples based on the results.

Information about your objectives and audience, along with the principles and tools described throughout this book, will allow you to tailor your approach, choosing terminology, analogies, formats, and a level of detail that best convey the purpose, findings, and implications of your study to the people who will read it. Throughout this book, I return often to issues about audience and objectives as they relate to specific aspects of writing about numbers.

A WRITER'S TOOLKIT

Writing about numbers is more than simply plunking a number or two into the middle of a sentence. You may want to provide a general image of a pattern, or you may need specific, detailed information. Sometimes you will be reporting a single number, other times many numbers. Just as a carpenter selects among different tools depending on the job, people who write about numbers have an array of tools and techniques to use for different purposes. Some approaches do not suit certain jobs, whether in carpentry (e.g., welding is not used to join pieces of wood) or in writing about numbers (e.g., a pie chart cannot be used to show trends). And just as there may be several appropriate tools for a task in carpentry (e.g., nails, screws, glue, or dowels to fasten together wooden components), in many instances any of several tools could be used to present numbers.

There are three basic tools in a writer's toolkit for presenting quantitative information: prose, tables, and charts.

Prose

Numbers can be presented as a couple of facts or as part of a detailed description of findings. A handful of numbers can be described in a sentence or two, whereas a complex statistical analysis can require a page or more. In the body of a paper, newspaper article, book, or blog, numbers are incorporated into full sentences. In slides for a speech, the executive summary of a report, chartbook pages, or on a research poster, numbers are often reported in a bulleted list, with short phrases used in place of complete sentences. Detailed background information is often given in footnotes (for a sentence or two) or appendixes (for longer descriptions).

Tables

Tables use a grid to present numbers in a predictable way, guided by labels and notes within the table. A simple table might present unemployment rates in each of several cities. A more complicated table might show relationships among three or more variables such as unemployment rates by city over a 20-year period, or results of statistical models analyzing unemployment rates. Tables are often used to organize a detailed set of numbers in appendixes, to supplement the information in the main body of the work.

Charts

There are pie charts, bar charts, line charts, scatter charts, and the many variants of each. Like tables, charts organize information into a predictable format: the axes, legend, and labels of a well-designed chart lead the audience through a systematic understanding of the patterns being presented. Charts can be simple and focused, such as a pie chart showing the racial composition of your study sample. Or they can be complex, such as a "high/low/close" chart illustrating stock market activity across a week or more.

As an experienced carpenter knows, even when any of several tools could be used for a job, often one of those options will work better in a specific situation. If there will be a lot of sideways force on a joint, glue will not hold well. If your listening audience has only 30 seconds to grasp a numerical relationship, a complicated table will be overwhelming. If kids will be playing floor hockey in your family room, heavy-duty laminated flooring will hold up better than parquet. If your audience needs many detailed numbers, a table will organize those numbers better than sentences.

With experience, you will learn to identify which tools are suited to different aspects of writing about numbers, and to choose among the workable options. Those of you who are new to writing about numbers can consider this book an introduction to carpentry—a way to familiarize yourself with the names and operations of each of the tools and the principles that guide their use. Those of you who have experience writing about numbers can consider this a course in advanced techniques, with suggestions for refining your approach and skills to communicate quantitative concepts and facts more clearly and systematically.

IDENTIFYING THE ROLE OF THE NUMBERS YOU USE

When writing about numbers, help your readers see where those numbers fit into the story you are telling—how they answer some question you have raised. A naked number sitting alone and uninterpreted is unlikely to accomplish its purpose. Start each paragraph with a topic sentence or thesis statement, then provide evidence that supports or refutes that statement. A short newspaper article on wages might report an average wage and a statistic on how many people earn the minimum wage. Longer, more analytic pieces might have several paragraphs or sections, each addressing a different question related to the main topic. A report on wage patterns might report overall wage levels, then examine how they vary by educational attainment, work experience, and other factors. Structure your paragraphs so your audience can follow how each section and each number contribute to the overall scheme.

To tell your story well, you, the writer, need to know *why* you are including a given fact or set of facts in your work. Think of the numbers as the answer to a word problem, then step back and identify (for yourself) and explain (to your readers) both the question and the answer. This approach is much more informative for readers than encountering a number without knowing why it is there. Once you have identified the objective and chosen the numbers, convey their purpose to your readers. Provide a context for the numbers by relating them to the issue at hand. Does a given statistic show how large or common something is? How small or infrequent? Do trend data illustrate stability or change? Do those numbers represent typical or unusual values? Often, numerical benchmarks such as thresholds, historical averages, highs, or lows can serve as useful contrasts to help your readers grasp your point more effectively: compare current average wages with the living wage needed to exceed the poverty level, for example.

ITERATIVE PROCESS IN WRITING

Writing about numbers is an iterative process. Initial choices of tools may later prove to be less effective than some alternative. A table layout might turn out to be too simple or too complicated, or you might conclude that a chart would be preferable. You might discover as you write a description of the patterns in a table that a different table layout would highlight the key findings more efficiently. You might need to condense a technical description of patterns for a research report into bulleted statements for an executive summary, or simplify them into charts for a speech or issue brief.

To increase your virtuosity at writing about numbers, I introduce a wide range of principles and tools to help you assess the most effective way to present your results. I encourage drafting tables and charts with pencil and paper before creating the computerized version, and outlining key findings before you describe a complex pattern, allowing you to separate the work into distinct steps. However, no amount of advance analysis and planning can envision the perfect final product, which likely will emerge only after several drafts and much review. Expect to have to revise your work, considering along the way the variants of how numbers can be presented.

OBJECTIVES OF THIS BOOK

How This Book Is Unique

Writing about numbers is a complex process: it involves finding pertinent data, identifying patterns, calculating comparisons, organizing ideas, designing tables or charts, and finally, writing prose. Each of these tasks alone can be challenging, particularly for novices. Adding to the difficulty is the final task of integrating the products of those steps into a coherent whole while keeping in mind the appropriate level of detail for your audience. Unfortunately, these steps are usually taught separately, each covered in a different book or course, discouraging authors from thinking holistically about the writing process.

This book integrates all of these facets into one volume, pointing out how each aspect of the process affects the others. For instance, the patterns in a table are easier to explain if that table was designed with both the statistics and writing in mind. An example will work better if the objective, audience, and data are considered together. By teaching all of these steps in a single book, I encourage you to consider both the "trees" (the tools, examples, and sentences) and the "forest" (your overall research question and its context). This approach will yield a clear, coherent story about your topic, with numbers playing a fundamental but unobtrusive role.

Another unique feature of this book is the "poor/better/best" teaching device that I developed to illustrate how to apply the various principles and skills. Many people find it challenging to apply new abstract ideas to specific situations—an essential step in writing about numbers. Willingham (2009) describes the importance of seeing new, abstract ideas in the context of things we already know, and points out that what we already know is concrete. To address the challenge of learning to master abstract ideas related to writing about numbers, I provide examples of how to apply a principle or skill such as "specify direction and magnitude" (chapter 2), illustrated with a concrete, familiar topic. I start by presenting a "poor" version of a prose description, table, or chart that did *not* follow that principle. I annotate that example to point out the specific aspects that were ineffective, along the way illustrating some of the most common errors I have observed when teaching that skill. I then provide "better" and "best" versions of that prose, table, or chart, annotated to explain why those versions represent improved applications of that principle.

The "poor" examples are adapted from ones I have encountered while writing and reviewing research papers and proposals, teaching research methods and writing courses, or attending and giving presentations to academic, policy, and business audiences. These examples may reflect lack of familiarity with quantitative concepts, poor writing or design skills, indoctrination into the jargon of a technical discipline, or failure to take the time to adapt materials for the intended audience and objectives. The principles and "better" examples will help you avoid similar pitfalls in your own work.

What This Book Is Not

Although this book deals with both writing and numbers, it is neither a writing manual nor a math or statistics book. Rather than restate principles that apply to other types of writing, I concentrate on those that are unique to writing about numbers and those that require some translation or additional explication. I assume a solid grounding in basic expository writing skills such as organizing ideas into a logical paragraph structure and using evidence to support a thesis statement. For good general guides to expository writing, see Strunk and White (1999) or Zinsser (1998). Other excellent resources include Lanham (2000) for revising prose and Montgomery (2003) for writing about science.

I also assume a good working knowledge of elementary quantitative concepts such as ratios, percentages, averages, and simple statistical tests, although I explain some mathematical and statistical issues along the way. See Kornegay (1999) for a dictionary of mathematical terms, Utts (1999) or Moore (1997) for good introductory guides to statistics, and Chambliss and Schutt (2012) or Lilienfeld and Stolley (1994) on study design. Those of you who write about multivariate analyses will benefit from the more advanced version of this book (Miller 2013a).

How This Book Is Organized

This book encompasses a wide range of material, from broad planning principles to specific technical details. The first part of the book, "Principles," lays the groundwork, describing a series of guidelines that form the basis for planning and evaluating your writing about numbers. The next part, "Tools," explains the nuts-and-bolts tasks of selecting, calculating, and presenting the numbers you will describe in your prose. The third part, "Pulling It All Together," demonstrates how to apply these principles and tools to write full papers, speeches, and other types of documents about numbers for both scientific and nonscientific audiences. For a study guide with problem sets and suggested course extensions as well as podcasts that help you apply ideas from this book to your own research or to courses that you teach, see https://www.press.uchicago.edu/books/miller.

Part I

PRINCIPLES

In this part, I introduce a series of fundamental principles for writing about numbers, ranging from setting the context to concepts of statistical significance to more technical issues such as types of variables, examining distributions, and using standards. These principles can be used to help you plan and evaluate elements of your numeric communications. Later parts of the book build on these principles and show you how to incorporate them into a complete paper or speech about an application of quantitative analysis.

12345678910111213

SEVEN BASIC PRINCIPLES

In this chapter, I introduce seven basic principles to increase the precision and power of your quantitative writing. I begin with the simplest, most general principles, several of which are equally applicable to other types of writing: setting the context; choosing simple, plausible examples; and defining your terms. Next, I introduce principles for choosing among prose, tables, and charts. Last, I cover several principles that are more specific to quantitative tasks: reporting and interpreting numbers, specifying direction and magnitude of associations, and summarizing patterns. I accompany each of these principles with illustrations of how to write (and how not to write) about numbers.

ESTABLISHING THE CONTEXT FOR YOUR FACTS
"The W's"
Context is essential for all types of writing. Few stories are told without somehow conveying "who, what, when, and where," or what journalists call "the W's." Without them your audience cannot interpret your numbers and will probably assume that your data describe everyone in the current time and place (e.g., the entire population of the United States in 2014). This unspoken convention may seem convenient. However, if your numbers are read later or in a different situation without information about their source, they can be misinterpreted. Don't expect your readers to keep track of when a report was issued to establish the date to which the facts pertain. Even using such tricks, all they can determine is that the information predated publication, which leaves a lot of room for error. If you encounter data

BOX 2.1. NAMED PERIODS AND COHORTS

Some time periods or cohorts are referred to by names such as "the Great Depression," "the post-war baby boom," or "Generation X," the dates varying from source to source. Generation X is loosely defined as the generation following the baby boom, but has been variously interpreted as "those born between 1965 and 1980," "those raised in the 1970s and 1980s," or even "those born since the mid-1960s" (scary, since it is lacking an end date, unless you look at when the article was published) (Jochim 1997). When reporting numbers about a "named" period for general background purposes, varying definitions probably don't matter very much. However, if your readers need precise comparisons, specify the range of dates. If you directly compare statistics from several sources, point out any variation in the definitions.

without the W's attached, either track down the associated contextual information and report it, or don't use those facts.

To include all of the W's, some beginners write separate sentences for each one, or write them in a stilted list: "The year was 2014. The place was the United States. The numbers reported include everyone of all ages, racial groups, and both sexes. [Then a sentence reporting the pertinent numbers]." Setting the context doesn't have to be lengthy or rote. In practice, each of the W's requires only a few words or a short phrase that can be easily incorporated into the sentence with the numbers. Suppose you want to include some mortality statistics in the introductory section of a paper about the Black Plague in fourteenth-century Europe.

> *Poor:* "There were 25 million deaths."
>
> *This statement lacks information about when and where these deaths occurred, or who was affected (e.g., certain age groups or occupations). It also fails to mention whether these deaths were from the Black Plague alone or whether other causes also contributed to that figure.*

Better: "During the fourteenth century, 25 million people died in Europe."

Although this statement specifies the time and place, it still does not clarify whether the deaths were from all causes or from the plague alone.

Best: "When the Black Plague hit Europe in the latter half of the fourteenth century, it took the lives of 25 million people, young and old, city dwellers and those living in the countryside. The disease killed about one-quarter of Europe's total population at the time (Mack, n.d.)."

This sentence clearly conveys the time, place, and attributes of the people affected by the plague, and provides information to convey the scale of that figure.

Despite the importance of specifying context, it is possible to take this principle too far: in an effort to make sure there is absolutely no confusion about context, some authors repeat the W's for every numeric fact. I have read papers that mention the date, place, and group literally in every sentence pertaining to numbers—a truly mind-numbing experience for both writer and reader. Ultimately, this obscures the meaning of the numbers because those endless W's clutter up the writing. To avert this problem, specify the context for the first number in a paragraph, then mention it again in that paragraph only if one or more aspects of the context change.

"When the Black Plague hit Europe in the latter half of the fourteenth century, it took the lives of 25 million people. The disease killed about one-quarter of Europe's total population at the time." [*Add*] "Smaller epidemics occurred from 1300 to 1600."

The last sentence mentions new dates but does not repeat the place or cause of death, implying that those aspects of the context remain the same as in the preceding sentences.

If you are writing a description of numeric patterns that spans several paragraphs, occasionally mention the W's again. For longer descriptions, this will occur naturally as the comparisons you make vary from one paragraph to the next. In a detailed analysis of the plague, you might compare mortality from the plague to mortality from other causes in the same time and place, mortality from the plague in other places or other times, and a benchmark statistic to help people relate to the magnitude of the plague's impact. Discuss each of these points in separate sentences or paragraphs, with introductory topic phrases or sentences stating the purpose and context of the comparison. Then incorporate the pertinent W's into the introductory sentence or the sentence reporting and comparing the numbers.

Units

An important aspect of "what" you are reporting is the units in which it was measured. There are different systems of measurement for virtually everything we quantify—distance, weight, volume, temperature, monetary value, and calendar time, to name a few. Although most Americans continue to be brought up with the British system of measurement (distance in feet and inches; weight in pounds and ounces; liquid volume in cups, pints, and gallons; temperature in degrees Fahrenheit), most other countries use the metric system (meters, grams, liters, and degrees Celsius, respectively). Different cultural and religious groups use many different monetary and calendar systems.

Scale of measurement also varies, so that population statistics may be reported in hundreds, thousands, millions, or even billions of people, according to whether one is discussing local, national, or international figures. Because of these variations, if the units of measurement are not reported along with a fact, a number alone is virtually useless, as you will see in some amusing examples in chapter 4, where I discuss this important principle in depth.

PICKING SIMPLE, PLAUSIBLE EXAMPLES

As accomplished speakers know, one strong intuitive example or analogy can go a long way toward helping your audience grasp quantitative concepts. If you can relate calories burned in a recommended exercise to how many extra cookies someone could eat, or translate a tax reduction into how many dollars a typical family would save, you will have given your readers a familiar basis of comparison for the numbers you report.

Most people don't routinely carry scales, measuring cups, or radar guns with them, so if you refer to dimensions such as weight, volume, or speed, provide visual or other analogies to explain particular values. In a presentation about estimating the number of people attending the 1995 Million Man March, Joel Best held up a newspaper page to portray the estimated area occupied by each person (3.6 square feet).[1] This device was especially effective because he was standing behind the page as he explained the concept, making it easy for his audience literally to see whether it was a reasonable approximation of the space he—an average-size adult—occupied.

The choice of a fitting example or analogy is often elusive. Finding one depends on both the audience and the specific objective of your example.

Objectives of Examples

Most examples are used to provide background information that establishes the importance of the topic, to compare findings with earlier ones, or to illustrate the implications of results. Your objectives will determine the choice of an example. For introductory information, a couple of numeric facts gleaned from another source usually will do. In the results section of a detailed scientific report, examples often come from your own analyses, and appropriate contrasts within your own data or comparisons with findings from other sources become critical issues. Below I outline a set of criteria to get you started thinking about how to choose effective examples for your own work.

The logic behind choosing numeric illustrations is similar to that for selecting excerpts of prose in an analysis of a literary work or case examples in a policy brief. Some examples are chosen to be representative of a broad theme, others to illustrate deviations or exceptions from a pattern. Make it clear whether an example you are writing about is typical or atypical, normative or extreme. Consider the following ways to describe annual temperature:

Poor: "In 2001, the average temperature in the New York City area was 56.3 degrees Fahrenheit."
From this sentence, you cannot tell whether 2001 was a typical year, unusually warm, or unusually cool.

Better: "In 2001, the average temperature in the New York
City area was 56.3 degrees Fahrenheit, 1.5 degrees above
normal."
*This version clarifies that 2001 was a warm year, as well as
reporting the average temperature.*
Best: "In 2001, the average temperature in the New York City
area was 56.3 degrees Fahrenheit, 1.5 degrees above normal,
making it the seventh warmest year on record for the area."
*This version points out not only that temperatures for 2001 were
above average, but also just how unusual that departure was.*

Principles for Choosing Examples

The two most important criteria for choosing effective examples are
simplicity and plausibility.

SIMPLICITY

The oft-repeated mnemonic KISS—"Keep It Simple, Stupid"—applies
to both the choice and explication of examples and analogies. Although the definition of "simple" will vary by audience and length of
the work, your job is to design and explain examples that are straightforward, concrete, and familiar (Willingham 2009). The fewer terms
you have to define along the way, and the fewer logical or arithmetic steps you have to walk your readers through, the easier it will be
for them to understand the example and its purpose. The immensity
of the Twin Towers was really driven home by equating the volume
of concrete used in those buildings to the amount needed to build a
sidewalk from New York City to Washington, DC (Glanz and Lipton
2002)—especially to a reader who had recently completed the three-
hour train ride between those cities.

PLAUSIBILITY

A comparison example must be plausible: the differences between
groups or changes across time must be feasible biologically, behavior-
ally, politically, or in whatever arena your topic fits. If you calculate
the beneficial effects of a 20-pound weight loss on chances of a heart
attack but the average dieter loses only 10 pounds, your projection
will not apply to most cases. If voters are unlikely to approve more

than a 0.7% increase in local property taxes, projecting the effects of a 1.0% increase will overestimate potential revenue.

This is an aspect of choosing examples that is ripe for abuse: advocates can artificially inflate apparent benefits (or understate liabilities) by using unrealistically large or small differences in their examples. For instance, sea salt aficionados tout the extra minerals it provides in comparison to those found in regular ol' supermarket salt (sodium chloride). Although sea salt does contain trace amounts of several minerals, closer examination reveals that you'd have to eat about a quarter pound of it to obtain the amount of iron found in a single grape (Wolke 2002). The fact that two pounds of sodium chloride can be fatal provides additional perspective on just how problematic a source of iron sea salt would be.

Other factors to consider include relevance, comparability, target audience, and how your examples are likely to be used, as well as a host of measurement issues. Because the choice of examples has many subtle nuances, I devote the whole of chapter 8 to additional guidelines.

SELECTING THE RIGHT TOOL FOR THE JOB

The main tools for presenting quantitative information—prose, tables, and charts—have different, albeit sometimes overlapping, advantages and disadvantages. Your choice of tools depends on several things, including how many numbers are to be presented, the amount of time your audience has to digest the information, the importance of precise versus approximate numeric values, and, as always, the nature of your audience. Chapters 6 and 7 provide detailed guidelines and examples. For now, a few basics.

How Many Numbers?

Some tools work best when only a few numbers are involved, others can handle and organize massive amounts of data. Suppose you are writing about how unemployment has varied by age group and region of the country in recent years. If you are reporting a few numbers to give a sense of the general pattern of unemployment for a short piece or an introduction to a longer work, a table or chart would probably be overkill. Instead, use a sentence or two:

"In February 2012, the unemployment rate for the United States was 8.3%, down from 9.0% a year earlier. Unemployment rates in each of the four major census regions also showed a minor decrease over that period (US Bureau of Labor Statistics 2012)."

If you need to include 10 years' worth of unemployment data on three age groups for each of four census regions, a table or chart is efficient and effective.

How Much Time?

When a presentation or memo must be brief, a chart, simple table, or series of bulleted phrases is often the quickest way of helping your audience understand your information. Avoid large, complicated tables: your audience won't grasp them in the limited time. For a memo or executive summary, write one bullet for each point in lieu of tables or charts.

Are Precise Values Important?

If in-depth analysis of specific numeric values is the point of your work, a detailed table is appropriate. For instance, if your readers need to see the fine detail of variation in unemployment rates over a decade or more, a table reporting those rates to the nearest tenth of a percentage point would be an appropriate choice. On the other hand, if your main objective is to show the general pattern of unemployment over that period, a chart would work better: all those numbers (and extra digits) in a table can distract and confuse.

"A chart is worth a thousand words," to play on the cliché. It can capture vast amounts of information far more succinctly than prose, and illustrate the size of a difference or the direction of a trend more powerfully than a sentence or a table. There is a tradeoff, however: it is difficult to ascertain exact values from a chart; avoid them if that is your objective.

Mixing Tools

In most situations, you will use a combination of tables, charts, and prose. Suppose you are writing a scholarly paper on unemployment patterns. You might include a few statistics on current unemployment rates in your introduction, a table to show how current unemployment varies by age group and region, and some charts to illustrate 10-year trends in unemployment by age group and region. To explain patterns in the tables or charts and relate them to the main purpose of the paper, describe those patterns in prose. For oral presentations, chartbooks, or automated slide shows, use bulleted phrases next to each chart or table to summarize the key points. Examples of these formats appear in later chapters.

As a general rule, don't duplicate information in both a table and a chart; you will only waste space and test your readers' patience. For instance, if I were to see both a table and a chart presenting unemployment rates for the same three age groups, four regions, and 10-year period, I would wonder whether I had missed some important point that one but not the other vehicle was trying to make. And I certainly wouldn't want to read the explanation of the same patterns twice—once for the table and again for the chart.

There are exceptions to every rule, and here are two. First, if both a quick sense of a general pattern *and* access to the full set of exact numbers matter, you might include a chart in the text, and a table in an appendix to report the detailed numbers from which the chart is constructed. Second, if you are presenting the same information to different audiences or in different formats, make both table and chart versions of the same data. You might use a table of unemployment statistics in a detailed academic journal article but show the chart in a presentation to your church's fund-raising committee for the homeless.

DEFINING YOUR TERMS (AND WATCHING FOR JARGON)
Why Define Terms?
Quantitative writing often uses technical language. To make sure your audience comprehends your information, define your terms, acronyms, and symbols.

UNFAMILIAR TERMS
Don't use phrases such as "opportunity cost" or "standardized mortality ratio" with readers who are unfamiliar with those terms. Ditto with abbreviations such as SES, LBW, or PSA. If you use technical

BOX 2.2. NAMES FOR NUMBERS

In addition to some of the more obvious jargon, other numeric terminology can confuse your audience. You might want to spice up your writing by using phrases such as "a century" instead of "100 years" or "the age of majority" instead of "age 18." Some of those phrases are widely understood, others a part of cultural literacy that depends on what culture you are from. That "a dozen" equals 12 is common knowledge in the United States, but the idea that "a baker's dozen" equals 13 is less universal. Writing for a modern American audience, I would hesitate to use terms such as "a fortnight" (14 nights or two weeks), "a stone" (14 pounds), or "a score" (20) without defining them. A British author or a historian could probably use them with less concern. Think carefully about using terms that require a pause (even a parenthetical pause) to define them, as it can interrupt the rhythm of your writing.

language without defining it first, you run the risk of intimidating an applied or lay audience and losing them from the outset. Or, if they try to figure out the meaning of new words or acronyms while you are speaking, they will miss what you are saying. If you don't define terms in written work, you either leave your readers in the dark, send them scurrying for a textbook or a dictionary, or encourage them to disregard your work.

TERMS THAT HAVE MORE THAN ONE MEANING

A more subtle problem occurs with words or abbreviations that have different meanings in other contexts. If you use a term that is defined differently in lay usage or in other fields, people may *think* they know what you are referring to when they actually have the wrong concept.

- To most people, a "significant difference" means a large one, rather than a difference that meets certain criteria for inferential statistical tests.[2] Because of the potential for confusion about the meaning of "significant," restrict its use to the statistical sense when describing statistical results. Many other adjectives such as "considerable," "appreciable," or even "big" can fill in ably to describe large differences between values.

- Depending on the academic discipline and type of analysis, the Greek symbol α (alpha) may denote the probability of Type I error, inter-item reliability, or the intercept in a regression model—three completely different concepts (Agresti and Finlay 1997).

- "Regression analysis" could mean an investigation into why Johnny started sucking his thumb again. Among statisticians, it refers to a technique for estimating the net effects of several variables on some outcome of interest, such as how diet affects child growth when illness and exercise are taken into account.

- The acronym PSA means "public service announcement" to people in communications, "prostate specific antigen" to health professionals, "professional services automation" in the business world, among more than 80 other definitions according to an online acronym finder.

These examples probably seem obvious now, but can catch you unaware. Often people become so familiar with how a term or symbol is used in a particular context that they forget that it could be confused with other meanings. Even relative newcomers to a field can become so immersed in their work that they no longer recognize certain terms as ones they would not have understood a few months before.

DIFFERENT TERMS FOR THE SAME CONCEPT
People from different fields of study sometimes use different terms for the same quantitative concept. For example, the term "scale" is sometimes referred to as "order of magnitude," and what some people call an "interaction" is known to others as "effect modification." Even with a quantitatively sophisticated audience, don't assume that people will know the equivalent vocabulary used in other fields. In 2002, the journal *Medical Care* published an article whose sole purpose was to compare statistical terminology across various disciplines involved in health services research, so that researchers could understand one

another (Maciejewski et al. 2002). After you define the term you plan to use, mention the synonyms from other fields represented in your audience to make sure they can relate your methods and findings to those from other disciplines.

To avoid confusion about terminology, scan your work for jargon before your audience sees it. Step back and put yourself in your readers' shoes, thinking carefully about whether they will be familiar with the quantitative terms, concepts, abbreviations, and notation. Show a draft of your work to someone who fits the profile of one of your future readers in terms of education, interest level, and likely use of the numbers, and ask them to flag anything they are unsure about. Then evaluate whether those potentially troublesome terms are necessary for the audience and objectives.

Do You Need Technical Terms?

One of the first decisions is whether quantitative terminology or mathematical symbols are appropriate for a particular audience and objective. For all but the most technical situations, *you* need to know the name and operation of the tools you are using to present numeric concepts, but your readers might not. When a carpenter builds a deck for your house, she doesn't need to name or explain to you how each of her tools works as long as she knows which tools suit the task and is adept at using them. You use the results of her work but don't need to understand the technical details of how it was accomplished.

To demonstrate their proficiency, some writers, particularly novices to scientific or other technical fields, are tempted to use only complex quantitative terms. However, some of the most brilliant and effective writers are so brilliant and effective precisely because they can make a complicated idea easy to grasp. Even for a quantitatively adept audience, a well-conceived restatement of a complex numeric relation underscores your familiarity with the concepts and enlightens those in the audience who are too embarrassed to ask for clarification.

WHEN TO AVOID JARGON ALTOGETHER
For nonscientific audiences or short pieces where a term would be used only once, avoid jargon altogether. There is little benefit to intro-

ducing new vocabulary or notation if you will not be using it again. And for nonstatisticians, equations full of Greek symbols, subscripts, and superscripts are more likely to reawaken math anxiety than to promote effective communication. The same logic applies to introductory or concluding sections of scientific papers: using a new word means that you must define it, which takes attention away from your main point. If you will not be using that term again, find other ways to describe numeric facts or patterns. Replace complex or unfamiliar words, acronyms, or mathematical symbols with their colloquial equivalents, and rephrase complicated concepts into more intuitive ones.

As an illustration, suppose an engineering firm has been asked to design a bridge between Littletown and Midville. To evaluate which materials last the longest, they use a statistical technique called failure time analysis. They are to present their recommendations to local officials, few of whom have technical or statistical training.

Poor: "The relative hazard of failure for material C was 0.78."
The key question—which material will last longer—is not
answered in ways that the audience will understand. Also, it is
not clear which material is the basis of comparison.
Better: "Under simulated conditions, the best-performing material
(material C) lasted 1.28 times as long as the next best choice
(material B)."
This version presents the information in terms the audience can
comprehend: how much longer the best-performing material
will last. Scientific jargon that hints at a complicated statistical
method has been translated into common, everyday language.
Best: "In conditions that mimic the weather and volume
and weight of traffic in Littletown and Midville, the best-
performing material (material C) has an average expected
lifetime of 64 years, compared with 50 years for the next best
choice (material B)."
In addition to avoiding statistical terminology related to failure
time analysis, this version gives specific estimates of how long
the materials can be expected to last, rather than just reporting

the comparison as a ratio. It also replaces "simulated conditions" with the particular issues involved—ideas that the audience can relate to.

WHEN TO USE AND PARAPHRASE JARGON

Many situations call for one or more scientific terms for numeric ideas. You might refer repeatedly to unfamiliar terminology. You might use a word or symbol that has several different meanings. You might refer to a concept that has different names in different disciplines. Finally, you might have tried to "explain around" the jargon, but discovered that explaining it in nontechnical language was too convoluted or confusing. In those instances, use the new term, then define or rephrase it in other, more commonly used language to clarify its meaning and interpretation. Suppose a journalist for a daily newspaper is asked to write an article about international variation in mortality.

> Poor: "In 2009, the crude death rate (CDR) for Sweden was 10 deaths per 1,000 people and the CDR for Ghana was 8 deaths per 1,000 people (World Bank 2012). You would think that Sweden—one of the most highly industrialized countries— would have lower mortality than Ghana—a less developed country. The reason is differences in the age structure, so I calculated life expectancy for each of the two countries. To calculate life expectancy, you apply age-specific death rates for every age group to a cohort of ... [You get the idea ...]. Calculated life expectancies for Sweden and Ghana were 81 years and 63 years."
> *This explanation includes a lot of background information and jargon that the average reader does not need, and the main point gets lost among all the details. Using many separate sentences, each with one fact or definition or calculation, also makes the presentation less effective.*
> Better (For a nontechnical audience): "In 2009, people in Ghana could expect to live until age 63 years, on average, compared to age 81 in Sweden. These life expectancies reflect much lower mortality rates in Sweden (World Bank 2012)."

This version conveys the main point about differences in mortality rates without the distracting detail about age distributions and how to calculate life expectancy.

Better (For a longer, more technical article): "In 2009, the crude death rate (CDR) for Sweden was 10 deaths per 1,000 people and the CDR for Ghana was 8 deaths per 1,000, giving the appearance of slightly more favorable survival chances in Ghana (World Bank 2012). However, Sweden has a much higher share of its population in the older age groups (18% aged 65 and older, compared to only 4% in Ghana), and older people have higher death rates. This difference pulls up the average death rate for Sweden. Life expectancy—a measure of mortality that corrects for differences in the age distribution—shows that in fact survival chances are much better in Sweden, where the average person can expect to live for 81 years, than in Ghana (63 years)."

This version conveys the main point about why life expectancy is the preferred measure and rephrases it in ways that introduce the underlying concepts: that older people have higher mortality, and that Sweden has a higher share of older people.

WHEN TO RELY ON TECHNICAL LANGUAGE

Although jargon can obscure quantitative information, equations and scientific phrasing are often useful, even necessary. When tradesmen talk to one another, using the specific technical names of their tools, supplies, and methods makes their communication more precise and efficient, which is the reason such terms exist. Being familiar with a "grade 8 hex cap bolt," they know immediately what supplies they need. A general term such as "a bolt" would omit important information. Likewise, if author and audience are proficient in the same field, the terminology of that discipline facilitates communication. If you are defending your doctoral dissertation in economics, using the salient terminology demonstrates that you are qualified to earn your PhD. And an equation with conventional symbols and abbreviations provides convenient shorthand for communicating statistical relationships, model specifications, and findings to audiences that are conversant with the pertinent notation.

Even for quantitatively sophisticated audiences, define what you mean by a given term, acronym, or symbol to avoid confusion among different possible definitions. I also suggest paraphrasing technical language in the introductory and concluding sections of a talk or paper, saving the heavy-duty jargon for the methodological and analytic portions. This approach reminds the audience of the purpose of the analyses, and places the findings back in a real-world context—both important parts of telling your story with numbers.

REPORTING *AND* INTERPRETING

Why Interpret?

Reporting the numbers you work with is an important first step toward effective writing about numbers. By including the numbers in the text, table, or chart, you give your readers the raw materials with which to make their own assessments. After reporting the raw numbers, interpret them. An isolated number that has not been introduced or explained leaves its explication entirely to your readers. Those who are not familiar with your topic are unlikely to know which comparisons to make or to have the information for those comparisons immediately at hand. To help them grasp the meaning of the numbers you report, provide the relevant data and explain the comparisons. Consider an introduction to a report on health care costs in the United States, where you want to illustrate why these expenditures are of concern.

> *Poor:* "In 2011, total expenditures on health care in the United States were estimated to be more than $2.7 trillion (Centers for Medicare and Medicaid Services 2013)."
> *From this sentence, it is difficult to assess whether total US expenditures on health care are high or low, stable or changing quickly. To most people, $2.7 trillion sounds like a lot of money, but a key question is "compared to what?" If they knew the total national budget, they could do a benchmark calculation, but you will make the point more directly if you do that calculation for them.*
> *Better:* "In 2011, total expenditures on health care in the United States were estimated to be more than $2.7 trillion, equivalent

to $8,658 for every man, woman, and child in the nation (Centers for Medicare and Medicaid Services 2013)."

This simple translation of total expenditures into a per capita figure takes a large number that is difficult for many people to fathom and converts it into something that they can relate to. Readers can compare that figure with their own bank balance or what they have spent on health care recently to assess the scale of national health care expenditures.

Best (To emphasize trend): "Between 2000 and 2011, the total costs of health care in the United States nearly doubled, from $1,377 billion to $2,693 billion (Centers for Medicare and Medicaid Services 2013; table 1). Over that same period, the share of gross domestic product (GDP) spent for health care increased from 13.7% to 17.7% (Organisation for Economic Co-operation and Development [OECD] 2013, figure 7.2.2)."

By discussing how health care expenditures have changed across time, this version points out that the expenditures have risen markedly in recent years. The sentence on share of GDP spent on health care shows that these expenditures comprise a substantial portion of the national budget—another useful benchmark.

Best (To put the United States in an international context): "In the United States, per capita health expenditures averaged $8,658 in 2011, equivalent to 17.7% of gross domestic product (GDP). The percentage of GDP spent on health care in the United States was 50% higher than the next highest country (the Netherlands, 11.9% of its GDP), and 90% higher than the OECD average of 9.3% (OECD 2013; figure 7.2.1)."

This description reveals that health care expenditures in the United States were the highest of any country, and reports how much higher compared to the next highest country and the average for a widely used international benchmark (the OECD). By using percentage of GDP as the measure, this comparison avoids the issue that countries with smaller populations would be expected to spend fewer total dollars but could still have higher per capita or percentage of GDP expenditures on health.

Why Report the Raw Numbers?

Although it is important to interpret quantitative information, it is also essential to report the numbers. If you *only* describe a ratio or percentage change, for example, you will have painted an incomplete picture. Suppose that a report by the local department of wildlife states that the density of the deer population in your town is 30% greater than it was five years ago but does not report the density for either year. A 30% difference is consistent with many possible combinations: 0.010 and 0.013 deer per square mile, or 5.0 and 6.5, or 1,000 and 1,300, for example. The first pair of numbers suggests a very sparse deer population, the last pair an extremely high concentration. Unless the densities themselves are mentioned, you can't determine whether the species has narrowly missed extinction or faces an over-population problem. Furthermore, you can't compare density figures from other times or places.

SPECIFYING DIRECTION AND MAGNITUDE OF AN ASSOCIATION

Writing about numbers often involves describing relationships between two or more variables. To interpret an association, explain both its shape and size rather than simply stating whether the variables are correlated.[3] Suppose an educational consulting firm is asked to compare the academic and physical development of students in two school districts, one of which offers a free breakfast program. If the consultants do their job well, they will report *which* district's students are bigger, faster, and smarter, as well as *how much* bigger, faster, and smarter.

Direction of Association

Variables can have a *positive* or *direct* association (as the value of one variable increases, the value of the other variable also increases) or a *negative* or *inverse* association (as one variable increases, the other decreases). Physical gas laws state that as the temperature of a confined gas rises, so does pressure; hence temperature and pressure are positively related. Conversely, as the price of a pair of jeans rises, the demand for jeans falls, so price and demand are inversely related.

For nominal variables such as gender, race, or religion that are classified into categories that have no inherent order, describe direction of association by specifying which category has the highest or lowest value (see chapter 4 for more about nominal variables, chapter 9 for more on prose descriptions of associations). "Religious group is negatively associated with smoking" cannot be interpreted. Instead, write "Mormons were least likely to smoke," and mention how other religious groups compare.

Size of Association

An association can be large (a given change in one variable is associated with a big change in the other variable) or small (a given change in one variable is associated with a small change in the other). A 15-minute daily increase in exercise might reduce body weight by five pounds per month or only one pound per month, depending on type of exercise, dietary intake, and other factors. If several factors each affect weight loss, knowing which make a big difference can help people decide how best to lose weight.

To see how these points improve a description of a pattern, consider the following variants of a description of the association between age and mortality. Note that describing direction and magnitude can be accomplished with short sentences and straightforward vocabulary.

Poor: "Mortality and age are correlated."
> *This sentence doesn't say whether age and mortality are positively or negatively related or how much mortality differs by age.*

Better: "As age increases, mortality increases."
> *Although this version specifies the direction of the association, the size of the mortality difference by age is still unclear.*

Best: "Among the elderly, mortality roughly doubles for each successive five-year age group."
> *This version explains both the direction and the magnitude of the age/mortality association.*

Specifying direction of an association can also strengthen statements of hypotheses: state which group is expected to have the more favorable outcome, not just that the characteristic and the outcome are expected to be related. "Persons receiving Medication A are expected to have fewer symptoms than those receiving a placebo" is more informative than "symptoms are expected to differ in the treatment and control groups." Typically, hypotheses do not include precise predictions about the size of differences between groups.

SUMMARIZING PATTERNS

The numbers you present, whether in text, tables, or charts, are meant to provide evidence about some issue or question. However, if you provide only a table or chart, you leave it to your readers to figure out for themselves what that evidence says. Instead, digest the patterns to help readers see the general relationship in the table or chart; in other words, answer the word problem.

When asked to summarize a table or chart, inexperienced writers often make one of two opposite mistakes: (1) they report every single number from the table or chart in the text, or (2) they pick a few arbitrary numbers to contrast in sentence form without considering whether those numbers represent an underlying general pattern. Neither approach adds much to the information presented in the table or chart, and both can confuse or mislead the audience. Paint the big picture, rather than reiterating all of the little details. If readers are interested in specific values within the pattern you describe, they can look them up in the accompanying table or chart.

Why Summarize?

Summarize to relate the evidence back to the substantive topic: Do housing prices change across time as would be expected based on changing economic conditions? Are there appreciable differences in housing prices across regions? Summarize broad patterns with a few simple statements instead of writing piecemeal about individual numbers or comparing many pairs of numbers. For example, when describing the pattern in figure 2.1, answering a question such as "are housing prices rising, falling, or remaining stable?" is much more instructive than responding to "what were housing prices in 1980, 1981, 1982 . . . 1999, 2000 in the Northeast?" or "how much did hous-

Median sales price of new single-family homes, by region, United States, 1980–2000

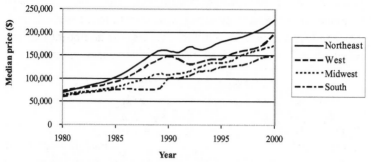

Figure 2.1. Generalizing patterns from a multiple-line trend chart
Source: US Census Bureau 2001a.

ing prices in the Northeast change between 1980 and 1981? Between 1981 and 1982? ..."

Generalization, Example, Exceptions—An Approach to Summarizing Numeric Patterns

Here is a mantra I devised to help guide you through the steps of writing an effective description of a pattern involving three or more numbers or series of numbers: "generalization, example, exceptions," or GEE for short. The idea is to identify and describe a pattern in general terms, give a representative example to illustrate that pattern, and then explain and illustrate any exceptions. This approach can also be used to compare results across different subgroups, time periods, or outcome variables (see "'Generalization, Example, Exceptions' Revisited" in chapter 9), or to synthesize findings or theories in previous literature (see "Literature Review" in chapter 11).

GENERALIZATION

For a generalization, come up with a description that characterizes a relationship among most, if not all, of the numbers. In figure 2.1, is the general trend in most regions rising, falling, or stable? Does one region consistently have the highest housing prices over the years? Start by describing one such pattern (e.g., trends in housing prices in the Northeast), then consider whether that pattern applies to the other regions as well. Or figure out which region had the highest housing prices in 1980 and see whether it is also the most expensive region in 1990 and 2000. If the pattern fits most of the time and most places, it is a generalization. For the few situations it doesn't fit, you have an *exception* (see below).

> "As shown in figure 2.1, the median price of a new single-family home followed a general upward trend in each of the four major census regions between 1980 and 2000. This trend was interrupted by a leveling off or even a decline in prices around 1990, after which prices resumed their upward trajectory. Throughout most of the period shown, the highest housing prices were found in the Northeast, followed by the West, Midwest, and South (US Census Bureau 2001a)."
>
> *This description depicts the approximate shape of the trend in housing prices (first two sentences), then explains how the four regions compare to one another in terms of relative price (last sentence). There are two generalizations: the first about how prices changed across time, the second about regional rank in terms of price. Readers are referred to the accompanying chart, which depicts the relationships, but no precise numbers have been reported yet.*

EXAMPLE

Having described your generalizable pattern in intuitive language, illustrate it with numbers from your table or chart. This step anchors your generalization to the specific numbers upon which it is based. It ties the prose and table or chart together. By reporting a few illustrative numbers, you implicitly show your readers where in the table or chart those numbers came from as well as the comparison involved. They can then test whether the pattern applies to other times, groups, or places using other data from the table or chart. Having written the above generalizations about figure 2.1, include sentences that incorporate examples from the chart into the description.

> (To follow the trend generalization): "For example, in the Northeast region, the median price of a new single-family

home rose from $69,500 in 1980 to $227,400 in 2000, more than a three-fold increase in price."

(To follow the across-region generalization): "In 2000, the median prices of new single-family homes were $227,400 in the Northeast, $196,400 in the West, $169,700 in the Midwest, and $148,000 in the South."

EXCEPTIONS

Life is complicated: rarely will anything be so consistent that a single general description will capture all relevant variation in your data. Tiny blips can usually be ignored, but if some parts of a table or chart depart substantially from your generalization, describe those departures. When portraying an exception, explain its overall shape and how it differs from the generalization you have described and illustrated in your preceding sentences. Is it higher or lower? By how much? If a trend, is it moving toward or away from the pattern you are contrasting it against? In other words, describe both direction and magnitude of the change or difference between the generalization and the exception. Finally, provide numeric examples from the table or chart to illustrate the exception.

"In three of the four regions, housing prices rose throughout the 1980s. In the South, however, housing prices did not begin to rise until 1990, after which they rose at approximately the same rate as in each of the other regions."

The first sentence describes a general pattern that characterizes most of the regions. The second sentence describes the exception and identifies the region to which it applies. Specific numeric examples to illustrate both the generalization and the exception could be added to this description.

Other types of exceptions include instances where prices in all four regions were rising but at a slower rate in some regions, or where prices rose over a sustained period in some regions but fell appreciably in others. In other words, an exception can occur in terms of magnitude (e.g., small versus large change over time) as well as in direction (e.g., rising versus falling, or higher versus lower) (see chapter 9). Because learning to identify and describe generalizations and exceptions can be difficult, in appendix A you will find additional pointers about recognizing and portraying patterns and organizing the ideas for a GEE into paragraphs, with step-by-step illustrations for several different tables and charts.

CHECKLIST FOR THE SEVEN BASIC PRINCIPLES

- Set the context for the numbers you present by specifying the W's: who, what, when, and where the data pertain to.
- Choose effective examples and analogies.
  Use simple, familiar examples that your audience will be able to understand and relate to.
  Select contrasts that are plausible under real-world circumstances.
- Choose vocabulary to suit your readers.
  Define terms and mention synonyms from related fields for statistical audiences.
  Replace jargon and mathematical symbols with colloquial language for nontechnical audiences.
- Decide whether to present numbers in text, tables, or figures.
  Determine how many numbers you need to report.
  Estimate how much time your audience has to grasp your data.
  Assess whether your readers need exact values.
- Report and interpret numbers in the text.
  Report them and specify their purpose.
  Interpret and relate them back to your main topic.
- Specify both the direction and size of an association between variables.
  If a trend, is it rising or falling, and how steeply?
  If a difference across groups or places, which has the higher value, and by how much?
- To describe a pattern involving many numbers, summarize the overall pattern rather than repeating all the numbers.
  Find a generalization that fits most of the data.
  Report a few illustrative numbers from the associated table or chart.
  Describe exceptions to the general pattern.

CAUSALITY, STATISTICAL SIGNIFICANCE, AND SUBSTANTIVE SIGNIFICANCE

A common task when writing about numbers is describing a relationship between two or more variables, such as the association between math curriculum and student performance, or the associations among diet, exercise, and heart attack risk. After portraying the shape and size of the association, interpret the relationship, assessing whether it is "significant" or "important."

Although many people initially believe that importance is based only on the size of an association—the bigger the difference across groups, the more important the association—in practice, this appraisal involves more thought. There are three key questions to consider. First, is the association merely a spurious correlation, or is there an underlying *causal* relationship between the variables? Second, is that association *statistically* significant? And third, is it *substantively* significant or meaningful? Only if all three criteria are satisfied does it make sense to base programs or policy on that association, seeking to improve student performance by changing math curriculums, for example. To provide a common basis for understanding these principles, below I review some needed statistical terms and concepts of study design, and provide references that treat these concepts in greater depth.

CAUSALITY

Many policy issues and applied scientific topics address questions such as, If we were to change *x*, would *y* get better? Will a new curriculum improve math comprehension and skills? Are e-cigarettes a gateway to smoking conventional cigarettes? If we dye white-haired people's hair some other color, will it increase their life spans? For permanent characteristics like gender, the question is slightly different: Is the difference across groups real, such that targeting cases based on those traits would be an appropriate strategy? Is it really gender that explains lower average earnings among women (implying gender discrimination), or are differences in work experience the cause?

Explanations for Associations

Anyone with even a passing acquaintance with statistics has probably heard the phrase "correlation does not necessarily mean causation." If an association between two variables x and y is statistically significant (see "Statistical Significance" below), that does not necessarily mean that x caused y or that y caused x. An association between two variables can be due to causality, confounding, bias, or simple happenstance.

CAUSAL ASSOCIATIONS

A causal association means that if the ostensible cause ("predictor" or "independent" variable) is changed, the hypothesized effect ("outcome" or "dependent" variable) will change in response. If a new curriculum is really the cause of better math performance, then adopting that curriculum should improve test scores. Establishing a plausible mechanism by which the first variable could affect the second helps build the case for a causal association. If you can show that the new math curriculum improves test scores by addressing previously neglected math facts or skills, that information strengthens the argument that the new curriculum is the reason for the better performance. To identify such mechanisms, know the theoretical context and base of existing knowledge for your topic.

Reverse causation occurs when what was believed to be the cause is actually the effect. For example, newspaper vendors are among the least healthy of all workers. How could that be? Selling newspapers doesn't seem like a very risky enterprise. It turns out that people who are too ill or disabled to perform other jobs are more likely than other people to become newspaper vendors because they are able to sell papers despite health problems that would prevent them from doing other jobs. Hence the ill health is what causes the occupation choice, not the reverse. To detect reverse causation, consider the time sequence of the two variables: which occurred first?

CONFOUNDING

If two variables are associated because of their mutual association with another variable, that relationship is *confounded* by that third variable (Abramson 1994). In some relationships affected by confounding, the third variable completely explains the association between the other two, in which case we say that association is *spurious*. People with white hair have higher mortality than people with other hair colors, but dyeing their hair black or blond is unlikely to improve their survival chances. Why? Because both the white hair and higher mortality are caused by old age (with some variation, of course), so the association between hair color and mortality is spurious rather than causal—it is wholly explained by the fact that both are associated with age.

In other associations, confounding is only a partial explanation: taking into account one or more confounding factors reduces the size of the association between a predictor and outcome, but the predictor retains some explanatory role. For example, both a high-fat diet and a lack of exercise are associated with higher risk of a heart attack, but diet and exercise are correlated. Thus when both are considered simultaneously, the size of the association between each predictor and heart attack risk is reduced; each of those variables confounds the relationship between the other predictor and heart attacks. Again, consider the theoretical and empirical context of your topic to assess whether confounding might be occurring.

BIAS

Bias is a systematic error in the observed patterns of one or more variables relative to their true values. In contrast to random error, where some measured values are higher than their actual values and some are lower, bias means that measurements consistently deviate from the true value in the same direction (Moore 1997). Bias can occur for several reasons, broadly classified into problems related to sampling and those related to measurement.

Sampling issues affect how cases are selected for a study. For instance, low-income families are less likely to have telephones. Thus studies that collect data using a telephone survey often underrepresent poor people, leading to biased estimates of knowledge, health, and other outcomes related to socioeconomic status. Another example: people who volunteer and qualify for a study are often different from nonparticipants. To be eligible for the Women's Health Initiative study of hormone replacement therapy (HRT), participants had to be free of heart disease at the start of the study, never have used HRT previously, and be willing to take HRT for several years—much longer than is typically prescribed (Kolata 2003). Hence the study findings might not apply to all women considering HRT. In these situations, the apparent association does not reflect the real pattern because the sample is not representative of the population of interest (see "Representativeness" in chapter 10).

Measurement problems relate to the ways data are collected, whether due to subjective reporting or bias in objective measurement. Respondents might shape their answers to more closely conform to social norms—playing down stigmatizing traits such as mental illness, or exaggerating desirable attributes like income, for example. Or an improperly calibrated scale might falsely add to the measured weights of every specimen.

As you interpret results, be alert to possible sampling or measurement biases that might cause the estimated association to differ from the pattern in the population from which your sample is drawn. See Chambliss and Schutt (2012) or Moore (1997) for more on bias in sampling and measurement.

Assessing Causality

How can you tell whether a relationship between two or more variables is causal? In the mid-1960s an advisory committee to the surgeon general agreed upon five criteria for establishing causality in epidemiologic studies (Morton, Hebel, and McCarter 2001). Four of those criteria are applicable to evaluating associations in other disciplines,[1] and similar principles have been established for the social sciences and other fields (Chambliss and Schutt 2012).

- *Consistency of association.* The association is observed in several different populations using different types of study design.
- *Strength of association.* A bigger difference in outcomes between cases with and without the purported causal factor indicates a stronger association.
- *Temporal relationship.* The cause preceded the effect. A correlation between two variables measured at the same time gives weaker evidence than one measuring the relationship between changes in the supposed cause and subsequent responses in the outcome.
- *Mechanism.* There is a plausible means by which the alleged cause could affect the outcome.

Much scientific research aims to assess whether relationships are causal, using empirical evidence as the basis of evaluation. Certain types of data are better for evaluating a potentially causal relationship. Data from a randomized controlled trial (experiment) or other "before and after" design provide more convincing evidence of a causal relationship than data where both the hypothesized cause and the hypothesized effect are measured at the same time ("cross-sectional" data). See Lilienfeld and Stolley (1994), Davis (1985), or Morton et al. (2001) for a thorough discussion of study design and causal inference.

Consider three approaches to evaluating whether a new math curriculum *causes* better math scores.

- An experimental study comparing math scores from schools with similar demographic and social makeup can provide convincing causal evidence. In an experiment, schools are randomly assigned either the new or the old math curriculum, using a coin toss, lottery, or random number table to decide which schools get which curriculum. Random assignment ensures that other possible causes of improved math scores are equally distributed among schools with both new and old curriculums, so those other factors cannot explain differences in observed test scores.

- Even in the absence of an experimental study, an improvement in math test scores after introduction of a new curriculum can lend support for the curriculum as the reason for better performance. The time sequence (temporal relationship) of curriculum adoption and math performance is unambiguous. However, other concurrent changes such as a decrease in class size or increase in teacher experience could be the reason for the better scores, and should be taken into account before inferring cause in either this or the preceding situation.
- A cross-sectional comparison of math scores for two schools, each of which happens to use one type of math curriculum, is less compelling because other factors that affect test scores could affect curriculum choice: an innovative school with involved parents or a larger budget might be more likely to adopt the new curriculum and to have better math scores regardless of curriculum, confounding the association between curriculum and math performance.

For situations in which random assignment isn't possible, "quasi-experimental" conditions are simulated by taking into consideration measures of possible confounding or mediating factors. For example, observational studies of different math curriculums compare performance differences by controlling statistically for other possible causal factors such as parental involvement and school budget (Allison 1999; Miller 2013a). Because it is impossible to measure all of the ways in which schools differ, however, evidence of causality from quasi-experimental studies is weaker than evidence from randomized experiments.

Causality as the Basis for Interventions
If confounding, bias, or reverse causality explains a correlation between variables, that association is not a good basis for policies or interventions aimed at changing the outcome. However, if a residual

association remains after you take confounding, bias, and reverse causation into account, then evaluate both the substantive and statistical significance of the remaining association to determine its policy relevance. For example, if exercise retains a large, statistically significant association with lowered heart attack risk even after the effect of diet has been considered, exercise programs could be an appropriate intervention against heart attacks.

Writing about Causality

How does knowledge about causality affect the way you write about relationships among the concepts under study? For analyses intended to inform policy or other interventions, convey whether the association is causal and describe possible causal mechanisms. Discuss alternative explanations such as bias, confounding, or reverse causation, indicating whether or not they are taken into account in your analysis. For noncausal relationships, explain the confounders, biases, or reverse causation that gave rise to the observed association.

VOCABULARY ISSUES

Carefully select the words you use to describe associations: verbs such as "affect" or "cause" and nouns such as "consequences" or "effects" all imply causality. "Correlated" or "associated" do not.

> *Poor*: "The effect of white hair on mortality was substantial, with five times the risk of any other hair color."
> *Poor* (version 2): "White hair increased the risk of dying by 400%."
> *The phrase "effect of x [white hair] on y [mortality]" implies causality. The active verb ("increased") also suggests that white hair brought about the higher mortality. Avoid such wording unless you have good reason to suspect a causal association.*
> *Slightly Better*: "The whiter the hair, the higher the mortality rate" or "As hair gets whiter, mortality increases."
> *These versions are written in neutral, purely descriptive language, failing to provide guidance about whether these patterns are thought to be causal.*

Better: "People with white hair had considerably higher mortality rates than people with a different color hair. However, most people with white hair were over age 60—a high-mortality age group—so the association between white hair and high mortality is probably due to their mutual association with older age."

Both the more neutral verb ("had") and linking both white hair and high mortality with old age help the audience grasp that white hair is not likely to be the cause of higher mortality. In this explanation, the focus is shifted from the attribute (white hair) to other possible differences between people who do and do not have the attribute(s) that could explain (confound) the hair color/mortality pattern.

Similar considerations apply to statements of hypotheses: phrase your statement to convey whether you believe the relationship to be causal or merely a correlation.

Poor: "We expect that the new math curriculum will be associated with higher math scores."

In the absence of a statement to the contrary, most readers will interpret this to mean that the new curriculum is expected to cause better math performance.

Better: "We expect that adoption of the new mathematics curriculum will improve math scores."

By using an active verb ("improve"), this statement explicitly conveys that the curriculum change is an expected cause of better math performance.

LIMITATIONS OF STUDY DESIGN FOR DEMONSTRATING CAUSALITY
While causality can be disproved by showing that even one of the causal criteria listed above is *not* true, it is much more difficult for one study to simultaneously show that all four criteria *are* true. It may be impossible to establish which event or characteristic came first, and often there are numerous potential unmeasured confounders and biases. For study designs that do not allow a cause-effect pattern to be tested well, point out those weaknesses and their implications for

inferring causality; see "Data and Methods in the Discussion Section" in chapter 10 for additional guidelines.

Box 3.1 is an excerpt from an article in the popular press that incorporates many of the above-mentioned aspects of how study design affects our ability to make causal interpretations of an association between two variables.

BOX 3.1. DISCUSSION OF HOW STUDY DESIGN AFFECTS CONCLUSIONS ABOUT CAUSALITY: ASSOCIATION BETWEEN USE OF E-CIGARETTES AND CONVENTIONAL CIGARETTES

"(1) Middle and high school students who used electronic cigarettes were more likely to smoke real cigarettes and less likely to quit than students who did not use the devices, a new study has found. They were also more likely to smoke heavily. (2) But experts are divided about what the findings mean. The study's lead author, Stanton Glantz, a professor of medicine at the University of California, San Francisco, who has been critical of the devices, said the results (A) suggested that the use of e-cigarettes was leading to less quitting, not more. 'The use of e-cigarettes does not discourage, and may encourage, conventional cigarette use among US adolescents,' the study concluded. . . . (B) But other experts said the data did not support that interpretation. They said that just because e-cigarettes are being used by youths who smoke more and have a harder time quitting does not mean that the devices themselves are the cause of those problems. It is just as possible, they said, that young people who use the devices were heavier smokers to begin with, or would have become heavy smokers anyway. . . .

"(3) Some experts worry that e-cigarettes are a gateway to smoking real cigarettes for young people, though most say the data is too skimpy to settle the issue. Others hope the devices could be a path to quitting. . . .

"(4) The new study drew on broad federal survey data from more than 17,000 middle school and high school students in 2011 and

---

BOX 3.1. (*CONTINUED*)

more than 22,000 in 2012. But (A) instead of following the same students over time—which many experts say is crucial to determine whether there has been a progression from e-cigarettes to actual smoking—the study (B) examined two different groups of students, essentially creating two snapshots....

"But David Abrams, executive director of the Schroeder Institute for Tobacco Research and Policy Studies at the Legacy Foundation, an antismoking research group, said the study's data do not support that conclusion. (5) 'I am quite certain that a survey would find that people who have used nicotine gum are much more likely to be smokers and to have trouble quitting, but that does not mean that gum is a gateway to smoking or makes it harder to quit,' he said. (6) He argued that there were many possible reasons that students who experimented with e-cigarettes were also heavier smokers—for example, living in a home where people smoke, belonging to a social circle where smoking is more common, or abusing drugs or alcohol." (Excerpted from Tavernise [2014])

COMMENTS

(1) Summarizes the main findings of the study, worded in cause-neutral language.

(2) Conveys two alternative interpretations of the observed association between e-cigarette and conventional cigarette use. Version (A) surmises that e-cigarettes are causally related to subsequent use of conventional cigarettes, as implied by the wording "leading to..." and the use of "discourage" and "encourage." Version (B) questions that interpretation, pointing out that association does not necessarily mean causation, and suggesting that the association between e- and conventional cigarette use is confounded by other characteristics such as those that predict heavy smoking.

(3) Restates causal interpretations, as suggested by the terms "gateway" and "path."

(4) Explains how the design of the study upon which the conclusions were based affected the capacity to test causality. (A) A prospective study following the same students over time to see which came first, e-cigarettes or conventional cigarettes, would provide strong evidence about whether the former caused the latter. Instead, the data (B) compared two cross-sections ("snapshots") comprising different sets of students, so it is impossible to determine the temporal order of the two smoking behaviors.

(5) Cites an example of an incorrect reverse causal interpretation of the observed association between nicotine gum and smoking as an analogy for how the association between e-cigarettes and conventional cigarettes could likewise be being interpreted "backward."

(6) Provides other, noncausal reasons why there might be an association between use of electronic and conventional cigarettes, citing several possible confounding factors.

STATISTICAL SIGNIFICANCE

Statistical significance is a formal way of assessing whether observed associations are likely to be explained by chance alone. It is an important consideration for most descriptions of associations between variables, particularly in scientific reports or papers, although some critics argue that its use is excessive and even incorrect (Ziliak and McCloskey 2008). In the absence of disclaimers about lack of statistical significance, readers tend to interpret reported associations as "real" and may use them to explain certain patterns or to generate solutions to problems. This is especially true if the association has already been shown to be causal.

In most instances, avoid a complicated discussion of the logic behind your statistical conclusions. Your statistics professor aside, many readers neither want nor need to hear how you arrived at your conclusions about statistical significance. Readers with statistical training will know how it was done if you name the statistical methods you used. Those without such training don't need the details. As in the carpenter analogy, the quality of the final product is affected by work done behind the scenes, but many consumers are interested only in that final product. It is up to you—the tradesperson—to ensure that appropriate steps were done correctly and to present the results in a neat, finished form.

An Aside on Descriptive and Inferential Statistics

As background for the remainder of this section, here's a quick review of the logic behind how statistical significance is assessed, to show how it relates to the other two aspects of "significance" discussed in this chapter. In chapters 9 and 13, I return to the topic of inferential statistics as I discuss how to present results of statistical tests to quantitatively-oriented and lay audiences, respectively. For a more thorough grounding in these concepts, consult Moore (1997), Utts (1999), or other introductory statistics textbooks.

Even if you have not studied statistics, you have probably encountered *descriptive statistics*, such as the mean, median, mode, and range in math scores among elementary school students. *Inferential statistics* take things a step further, testing whether differences between groups are statistically significant—not likely to occur by chance alone if there were no real association—as a way of formally comparing average math scores between schools, for example.

Inferential statistical tests evaluate how likely it would be to obtain the observed difference or a larger difference between groups under the assumption there is no difference. In statistical lingo, the assumption of "no difference" is called the *null hypothesis*. In the math example, the null hypothesis would state that average test scores for schools using the new and old math curriculums are equal. Most studies are based on a sample of all possible cases; thus random error affects estimates of average test scores and must be taken into account when assessing differences between groups. For example, inferences about the benefits of the math curriculum might be based on comparison of scores from a random sample of students rather than from all students following those curriculums. Because of random variation, the average scores in each of the curriculum groups

How Statistical Significance (or Lack Thereof) Affects Your Writing

How should you write about results of statistical tests? The answer depends on whether findings are statistically significant, your audience, the length and detail of the work, and the section of the paper.

STATISTICALLY SIGNIFICANT RESULTS

Many academic journals specify that you use a table to report statistical test results for all variables, but then limit your text description to only those results that are statistically significant, in most cases using $p < 0.05$ as the criterion. The $p < 0.05$ rule of thumb also applies for lay audiences, although you will use and present the statistical information differently; see "Statistical Significance" in chapter 13. Emphasizing statistically significant findings is especially important if you are investigating several different factors, such as how gender, race, class size, teacher's experience, and teacher's major field of study each affected students' math test scores. If only some traits associated with math scores are statistically significant, focus your discussion on those rather than giving equal prominence to all factors. This approach helps readers answer the main question behind the analysis: which factors can help improve math performance the most?

WHEN TO REPORT RESULTS THAT ARE NOT
STATISTICALLY SIGNIFICANT

The $p < 0.05$ rule notwithstanding, a nonstatistically significant finding can be highly salient if that result pertains to the main variable you are investigating: if the lack of statistical significance runs contrary to theory or previously reported results, report the numbers, size of the association, and the lack of statistical significance. In such situations, the lack of a difference between groups answers a key question in the analysis, so highlight it in the concluding section of the work (see "Numeric Information in a Concluding Section" in chapter 11).

Suppose earlier studies showed that students in School A were more likely to pass a standardized math test than students in School B. After a new math curriculum was implemented in School B, you find no difference between math scores in the two schools, or that the observed difference is not statistically significant. Mention the change in both size and statistical significance of the difference between the schools' scores compared to before the curriculum change, then explicitly relate the new finding back to theoretical expectations and results of previous studies. This approach applies to a short general-interest article or a policy brief as well as to scientific reports or papers.

"BORDERLINE" STATISTICAL SIGNIFICANCE

A controversial issue is what to do if the p-value is only slightly above 0.05, say $p = 0.06$ or $p = 0.08$. Such values fall in a gray area: strictly speaking they are not statistically significant according to conventional criteria, but they seem oh-so-close that they are difficult to ignore. How you handle such findings depends on several factors that influence statistical tests:

- The effect size
- The sample size
- The value of the test statistic and its associated p-value

If the effect size (e.g., the difference in math test scores between two schools, or correlation between exercise and heart attack risk) is very small, a larger number of cases is unlikely to increase the statistical significance of the association. Such associations are unlikely to be of substantive interest even if they are real and causal (see "Substantive Significance" below), so treat them as if they were not statistically significant. On the other hand, if the effect size is moderate to large, the p-value is in the gray area between $p < 0.05$ and $p < 0.10$, and the sample size is small, report the effect and its p-value, and mention the small sample size and its effect on the standard error. Unfortunately, all of these criteria are subjective (What is a "moderate effect size?" A "small sample size?") and opinions vary by discipline, so learn the conventions used in your field.

Writing about Statistical Significance

Tailor your description of statistical significance to meet the needs and abilities of your audience.

For most nontechnical audiences, keep discussion of statistical significance very simple or omit it entirely. No matter how carefully you try to phrase it, a discussion of the purpose and interpretation of statistical tests may confuse readers who are not trained in statistics. Instead of reporting p-values or test statistics, use statistical tests and consideration of causality as screens for what you report and how you discuss findings, or paraphrase the concepts behind the statistics into everyday language. See chapter 13 for examples.

SCIENTIFIC AUDIENCES

Statistically proficient readers expect a description of statistical results, but statistical significance is discussed differently in different parts of scientific papers. A typical scientific paper is organized into an introduction, data and methods section, results section, and discussion and conclusions (see chapters 10 and 11 for additional examples).

Results sections

In the results section of a quantitative piece, report the findings of statistical tests along with descriptors of the size and shape of relationships, following the guidelines in chapters 9 and 11. Indicators of statistical significance are expected features, whether in tables, text, or charts. Report exact p-values to two decimal places, writing "$p < 0.01$" for smaller values (see "Choosing a Fitting Number of Digits and Decimal Places" in chapter 4). Standard errors can be reported in a table. In the prose description, emphasize findings for the key variables in your research question, then limit description of results for other variables to those that were statistically significant, or follow the discussion of statistically significant findings with a simple list of those that were not.

> "On a mathematics test given to fourth graders in fall 2011, students in School A achieved a lower average score than students in School B (62.7% and 72.0% correct, respectively; $p < 0.001$)."

This description explicitly mentions the result of the statistical test (in this case, the p-value), along with the average scores.

Alternatively, express statistical significance using a confidence interval:

"On a mathematics test given to fourth graders in fall 2011, school A's average scores were statistically significantly lower than those in School B [as demonstrated by the fact that the respective confidence intervals do not overlap]. Students in School A scored on average 62.7% of questions correct (95% confidence interval [CI]: 58.5% to 66.9%), as against an average of 72.0% correct in School B (95% CI: 70.0% to 74.0%)."
By reporting the confidence intervals around each estimate, this version gives a range of substantive values for each school's average score, and allows readers to assess how the two schools' ranges compare. Include the phrase in brackets only for readers who are not familiar with how to interpret confidence intervals. Complement the written description with a chart showing the 95% CI around the estimated average scores for each school (see figure 7.11).

For findings that are not statistically significant:

"Math performance did not differ according to race, gender, or class size."
In a results section of a scientific paper, wording such as "did not differ" implies lack of a statistically significant difference. Report the math scores for each group and the associated test statistics in a table or chart, either with the text (for a longer report) or in an appendix.

For borderline cases such as when the p-value is slightly above 0.05 and sample sizes are small:

"Math scores for fall 2011 revealed no statistically significant difference between Schools A and B (67.7% and 72.0% correct, respectively; $p < 0.09$). However, only 20 students were tested in each school, hence standard errors were relatively large.

With a larger number of cases, the observed effect might have reached conventional levels of statistical significance."
By reporting the p-value, this version shows that the difference in test scores narrowly missed statistical significance, then explains the effect of small sample size on the standard error.

Treat an equivalent interscholastic difference in scores and border-line p-value based on a large sample size as a nonsignificant finding, without discussion of effect size or sample size.

Discussion sections

For a quantitative article or report, summarize key findings in the discussion and conclusions section, then relate them back to the research question. Discuss both statistically significant and nonsignificant results related to your main research question but don't reiterate nonsignificant findings for less central variables. To express the implications of your results, compare your major findings to theoretical expectations and previous empirical results in terms of size and statistical significance. Differences you discuss will typically be assumed to be statistically significant unless you state otherwise. In the discussion, describe statistical significance (or lack thereof) in terms of concepts and conclusions, using words rather than numeric results of statistical tests.

"An experimental study of the new math curriculum showed marked improvement in average test scores among students in high-income districts. In low-income districts, however, effects of the new curriculum were smaller and not statistically significant. Smaller class sizes were also associated with higher math scores, but did not explain the income difference in the effect of the new curriculum. Racial composition of the schools did not affect math performance."
The wording of this summary for the discussion section of a scientific paper is similar to that for a lay audience, but explicitly mentions statistical significance. Numeric results of statistical tests reported in the results section are not repeated in the discussion. To reiterate the importance of study design, the type

of study is mentioned but without the nitty-gritty details from the data and methods section about sample size, data collection, and so forth.

In all but the briefest articles, complete the picture by discussing possible reasons for discrepant findings about statistical significance from other studies, such as differences in the date or location of the study, the characteristics of the students involved, or how the data were collected (see "Data and Methods in the Discussion Section" in chapter 10). Similar principles apply to discussion of previous studies: when assessing the state of knowledge on a topic, discuss only statistically significant findings as "real."

SUBSTANTIVE SIGNIFICANCE

The third aspect of the "importance" of a finding is whether the size of an association is substantively significant, which is just a fancy way of asking "So what?" or "How much does it matter?" Is the cholesterol reduction associated with eating oatmeal large enough to be clinically meaningful? Is the improvement in math performance large enough to justify the cost of adopting the new curriculum? Statistical significance alone is not a good basis for evaluating the importance of some difference or change. With a large enough sample size, even truly microscopic differences can be statistically significant. However, tiny differences are unlikely to be meaningful in a practical sense. If every fourth grader in the United States were included in a comparison of two different math curriculums, a difference of even half a point in average test scores might be statistically significant because the sample size was so large. Is it worth incurring the cost of producing and distributing the new materials and training many teachers in the new curriculum for such a small improvement?

To assess the substantive importance of an association, place it in perspective by providing evidence about how that half-point improvement translates into mastery of specific skills, the chances of being promoted to the next grade level, or some other real-world outcome to evaluate whether that change is worthwhile. Report and evaluate both the prevalence and consequences of a problem: preventing a rare but serious health risk factor might be less beneficial than preventing a more common, less serious risk factor. For scientific audiences, consider including attributable risk calculations (chapter 5), cost-effectiveness analysis (e.g., Gold et al. 1996), or other methods of quantifying the net impact of proposed health treatments, policies, or other interventions as a final step in the results section. For other audiences, integrate results of those computations into the discussion and conclusions.

Address substantive significance in the discussion section as you consider what your results mean in terms of your original research question.

Poor: "The association between math curriculums and test scores was not very substantively significant."
Most people won't know what "substantively significant" means. In addition, this version omits both the direction and size of the association, and doesn't help readers assess whether the change is big enough to matter.

Better (for a scientific audience): "Although the improvement in math test scores associated with the new math curriculum is highly statistically significant, the change is substantively inconsequential, especially when the costs are considered: the half-point average increase in math scores corresponds to very few additional students achieving a passing score, or mastering important fourth-grade math skills such as multiplication or division. Spending the estimated $40 million needed to implement the new curriculum on reducing class sizes would likely yield greater improvements."
This description puts the results back in the context of the original research question. Both substantive and statistical significance are explicitly mentioned; causality is implicitly addressed using words such as "change" and "increase."

RELATIONS AMONG STATISTICAL SIGNIFICANCE, SUBSTANTIVE SIGNIFICANCE, AND CAUSALITY

The implications of a numeric association depend on whether that association is causal, statistically significant, and substantively meaningful. All three conditions are often necessary, and none alone may be sufficient to guarantee the importance of the association. As part of an ongoing debate in the social science literature about these issues, Ziliak and McCloskey (2004) showed that that over 80% of articles published in a top economics journal during the 1990s failed to distinguish between statistical and economic significance. Thompson (2004) points out that the failure to distinguish between substantive importance and statistical significance has a long history among authors from a variety of disciplines including psychology, sociology, education, medicine, and public health.

To avert such problems, avoid a long discussion of how much something matters substantively if the association is not causal or the difference between groups is not statistically significant. In scientific papers, review the evidence for statistical and substantive significance and explain that those two perspectives have distinctly different purposes and interpretations, devoting a separate sentence or paragraph to each (Miller and Rodgers 2008).

As you evaluate these criteria for the associations you are writing about, remember that even if one condition is satisfied, the others may not be.

- In nonexperimental studies, a statistically significant association does not necessarily mean causation: white hair and high mortality could be correlated 0.99 with a $p < 0.001$, but that does not make white hair a cause of high mortality. In experiments, where cases are randomized into treatment and control groups, however, confounding is usually ruled out, and statistically significant findings are often interpreted as causal.
- Conversely, a causal relationship does not necessarily guarantee statistical significance: e.g., random error or bias can obscure effects of a curriculum change on math performance.
- Statistical significance does not necessarily translate into substantive importance: the new math curriculum could be statistically significantly associated with math performance at $p < 0.05$, but the increment to math scores might be very small.

- Conversely, substantive importance does not ensure statistical significance: a large effect might not be statistically significant due to wide variation in the data or a small sample size.
- Causality does not automatically mean substantive importance: the new curriculum might improve math scores, but that change could be so slight as to be unworthy of investment.
- Substantive importance (a "big effect") does not automatically translate into causality, as in the white hair example.

CHECKLIST FOR CAUSALITY, STATISTICAL SIGNIFICANCE, AND SUBSTANTIVE SIGNIFICANCE

As you write about associations, discuss each of the following criteria as they pertain to your research question:

- Causality
  - Describe possible mechanisms linking the hypothesized cause with the presumed effect.
  - Discuss and weigh the merits of other, noncausal explanations, identifying sources of bias, confounding, or reverse causation.
  - Explain the extent to which your statistical methods overcome drawbacks associated with observational (nonexperimental) data, if applicable.
- Statistical significance
  - For a scientific audience,
    - report statistical significance in the results section (see chapters 9 and 11 for illustrative wording), mentioning p-values or confidence intervals in the text and reporting other types of statistical test results in a table;
    - return to statistical significance in the discussion, especially if findings are new or run counter to theory or previous studies. Restate findings in words, not numeric test statistics or p-values.
  - For a nontechnical audience, use the results of statistical tests to decide which findings to emphasize, but don't report the numeric details of the test results (see "Statistical Significance" in chapter 13).

- Substantive significance

    Evaluate the substantive importance of a finding by
    translating it into familiar concepts such as overall cost
    (or cost savings) or improvements in specific skills or other
    outcomes related to the topic at hand.

    For a technical audience, consider quantifying the difference
    using z-scores (see chapter 5).

- Integration of the three elements

    In your discussion or closing section, relate the findings back
    to the main research question, considering causality,
    statistical significance, and substantive significance
    together.

    For an observed numeric association to be used as the basis
    for a policy or intervention to improve the outcome under
    study, all three of those criteria should be satisfied.

1 2 3 **4** 5 6 7 8 9 10 11 12 13

FIVE MORE TECHNICAL PRINCIPLES

In addition to the principles covered in the previous two chapters, there are a handful of more technical issues to keep in mind as you write about numbers: understanding the types of variables you're working with, specifying units, examining the distributions of your variables, and finding standards against which to compare your data. These are important background steps before you decide which kinds of calculations, tables, and charts are appropriate, or select suitable values to contrast with one another. The final principle—choosing how many digits and decimal places to include—may seem trivial or merely cosmetic. However, measurement issues and ease of communication dictate that you select a level of numeric detail that fits your data and your audience.

UNDERSTANDING TYPES OF VARIABLES

Information about types of variables, also known as level of measurement (Chambliss and Schutt 2012), affects your choice of comparisons. Some calculations make sense only for variables measured in continuous units, others only for those classified into categories. You cannot calculate a mean test score if those data were collected as "pass/fail." Some variables allow only one response for each case, others permit more than one response. You cannot analyze patterns of multiple insurance coverage if each person was allowed to report only one type.

Choice of tools for presenting numbers also is affected by level of measurement and number of responses. In chapter 7, I explain why line or scatter charts are suitable for some types of variables, and bar or pie charts for others, and why some charts accommodate multiple responses per case while others work for only single-response variables. In chapter 9, I show how to describe distributions for different types of variables. For now, I introduce the vocabulary and properties of each kind of variable.

There are two main characteristics of each variable to consider: Was it measured in continuous or categorical fashion? And was each case allowed only one response or several responses?

Continuous and Categorical Variables

The type of variable—continuous or categorical—affects a variety of issues related to which types of quantitative comparisons are appropriate and how to interpret the results, described in chapter 5. Below I give basic definitions of these types of variables and how they relate to measurement issues and mathematical consistency checks.

CONTINUOUS VARIABLES

Continuous variables are measured in units such as years (e.g., age or date), inches (e.g., height or distance), or dollars (e.g., income or price), including those with decimal or fractional values. Their values are *not* grouped into ranges. Continuous variables are sometimes referred to as "quantitative variables," and come in one of two types: *interval* and *ratio* (Chambliss and Schutt 2012).

Interval variables can take on positive (> 0), zero, and negative (< 0) values. They can be compared using subtraction (calculating the "difference" or interval between values; see chapter 5) but not division (calculating the ratio of two values). Temperature as we conventionally measure it (whether Fahrenheit or Celsius)[1] is an interval variable: if it was 30°F yesterday and 60°F today, it makes sense to say that it is 30 degrees hotter today, but not that it is twice as hot. And using a ratio to compare temperatures above and below zero (e.g., −20°F versus +20°F) would be truly misleading.

Ratio variables can be compared using either subtraction or division because a value of zero can be interpreted as the lowest possible value. Distance is an example of a ratio variable: if it is two miles from your house to the mini-mart but four miles to the supermarket, you could either say the supermarket is "two miles farther than" or "twice as far as" the mini-mart.

CATEGORICAL VARIABLES

Categorical variables classify information into categories such as gender, race, or income group. They come in two flavors: *ordinal* and *nominal*. Ordinal ("ordered") variables have categories that can be ranked according to the values of those categories. A classic example is letter grades (A, B+, etc.). Income grouped into ranges of several thousand dollars is another ordinal variable, and an example of one for which units must also be specified. Likert-type items are another common type of ordinal variable, and are often used to measure the extent of a person's beliefs, attitudes, or feelings toward some topic (Chambliss and Schutt 2012). Example Likert items include:

> "Rate the extent of your agreement with the following statement: 'Automatic weapons should be banned.' Strongly disagree, disagree, neutral, agree, strongly agree."
>
> "How would you rate your health: 'Excellent, very good, good, fair, or poor?'"

Nominal ("named") variables, also referred to as "qualitative" variables, include gender, race, or religion, for which the categories have no inherent order.[2]

Continuous and categorical variables are not completely different animals. Continuous variables can be classified into ordered categories. You can create a variable "age group" out of a variable measuring age in years: a nine-year-old would be classified as a child. However you can't create a detailed continuous age variable from a categorical variable that encompasses several values in each age group: knowing that someone is a child does not tell you whether he is one or nine or eleven years old. Categorical versions of continuous variables are useful for simplifying information (by grouping age into five- or 10-year age groups, for example), or for indicating whether values of a continuous variable fall above or below a cutoff like retirement age.

DEFINING SENSIBLE CATEGORIES

Creating good categorical variables takes careful thought. If each case can have only one valid value, then every case should fall into exactly one group—no more, no less. Each person in the US is either US born or foreign born, and has only one age and one total income at a time.

In addition, every person has a birthplace, age, and income (even if it is zero or negative). In set theory, classifications such as these are known as *mutually exclusive* (nonoverlapping) and *exhaustive* (encompassing all possible responses). "Under 18 years," "18–64," and "65 and older" are mutually exclusive and exhaustive age groups because the youngest age group starts at zero and the oldest has no upper age limit, covering the full range of relevant answers.

Although mutually exclusive and exhaustive seem like straightforward concepts, they can be difficult to implement. Through 1990, the US Census question on race allowed each respondent to mark only one answer. For people who considered themselves to be multiracial, however, marking one race was not sufficient—the race categories weren't mutually exclusive. To resolve this issue, starting with the 2000 US Census, respondents were permitted to classify themselves as more than one race, and tabulations of race include both single- and multiple-race categories. These new categories allow each person to be counted once and only once. In 2000, multiple-race individuals accounted for about 2.4% of the total population (US Census Bureau 2002a).

A second problem arose for people of Hispanic origin: the Census Bureau treats Hispanic origin and race as separate characteristics covered in separate questions. According to Census Bureau definitions, Hispanic persons can be of any race, and persons of any race can be of Hispanic origin. However, many people of Hispanic origin consider their race to be Hispanic, not white, black, Asian, or Native American, so they often left the race question blank or checked "other" (Navarro 2003). For them, the list of categories was incomplete—in other words, it was not exhaustive.

"Other"

A common practice when collecting or classifying data is to create a category for "other" responses, allowing for answers researchers didn't anticipate. A survey question about the primary reason someone did not seek prenatal care might list cost, transportation issues, and the belief that pregnancy is a healthy condition, but overlook child care issues and language barriers, for example. An "other" category also permits researchers to later combine uncommon responses instead of creating separate categories for reasons mentioned by only a small share of respondents. Everyone's response fits somewhere, but there needn't be dozens of tiny categories to record and present in every table or chart.

If "other" encompasses a large share of cases, however, it can obscure important information, as when many Hispanics mark "other" for race on the census. Sometimes the response "none of the above" is used to indicate "other," but that approach misses the opportunity to find out what answer *does* apply. To avoid this problem, often questionnaires include a blank for respondents to specify what "other" means in their case. If many respondents list similar answers, they can be grouped into a new category for analysis or future data collection.

Missing values

If you've ever collected data, you know that people sometimes skip questions, refuse to respond, write something illegible or inappropriate, or mark more answers than are allowed. To account for all cases, an "unknown" or "missing" category is used to tabulate the frequency of missing information—an important fact when examining the distribution of your variables (see section below) or assessing whether your data are representative of the population under study (chapter 10).

"Not applicable" and missing by design

Some questions pertain only to a subset of cases. In a survey asking whether someone has changed residences recently and if so when, the "when" question does not apply to those who did not move. To make sure every case is accounted for, such cases are classified "not applicable." Differentiating "not applicable" from "missing" makes it easier for cases to be omitted from calculations that don't pertain to them, leaving out those who didn't move from an analysis of timing of moves rather than incorrectly lumping them in with people who did move but neglected to report when, for example. In some studies, certain questions are asked only of some participants. See "Missing by Design" in chapter 10 for details. Most data sets will designate separate numeric (or sometimes alphanumeric) codes for each of these reasons so users can distinguish among them, e.g., 97 = not applicable; 98 = module not administered to case; and 99 = item nonresponse.

Single- versus Multiple-Response Questions

For characteristics like current age or gender, each case has only one value. Other situations call for more than one answer per respondent. In a school board election, each voter might be asked to select three candidates, or a survey might ask respondents to list all types of health insurance coverage within their families, for example. The number of possible responses does not determine the type of variable—both single- and multiple-response items can be either continuous or categorical.

For many multiple-response questions, some people mark no responses, others mark several, and a few mark all possible answers. In some families, everyone has the same kind of health insurance, such as employer-sponsored or Medicaid. For those families, one response characterizes their insurance coverage. Some families have several kinds of health insurance, such as both employer-sponsored and Medicare, or both Medicaid and Medicare. Two or three responses are needed to characterize their coverage. Families that lack health insurance do not fit any categories. (On some surveys, "uninsured" might be listed as an option, in which case every family would have at least one response.)

Why does it matter that the number of answers each respondent can give to a question can vary from none to many? Because the principles of "mutually exclusive" and "exhaustive" don't apply to variables created from those questions. Having employer-sponsored health insurance does not preclude a family from also having Medicare for its elderly or disabled members, for example. If you tally up the total number of responses in all categories, they will exceed the number of families surveyed. For cases where none of the responses apply, the total number of responses can be less than the number of families surveyed. Consequently, the kinds of mathematical consistency checks used to evaluate the distribution of a single-response variable cannot be used for multiple-response questions. For single-response questions with an exhaustive set of response categories, the frequencies of all responses will add up to 100% of the cases. For multiple-response questions, that total could range from 0% (if no one marked any of the categories) to several hundred percent (if many people marked several categories).

Indexes and Scales

Indexes and scales[3] are composite variables designed to create a single reliable and valid measure of an underlying construct such as depression or support for abortion for which there is no direct measure (Treiman 2009; Chambliss and Schutt 2012). They are calculated from a series of items (questions) each of which measures a different dimension of that concept such as different depression symptoms. The categorical items to be used in an scale often share a consistent scoring scheme, e.g., either "yes/no," "correct/incorrect," or a series of Likert-type responses.

One way to compute a scale is by summing or averaging the scores (numeric codes) of the responses to individual items, resulting in a measure that is treated as a continuous variable. For instance, the Center for Epidemiological Studies of Depression (CESD) scale is calculated by summing the scores on 20 questions about frequency of specified depression symptoms, each coded as an ordinal variable from 0 ("not at all to less than one day/week") to 3 ("nearly every day") (Radloff and Locke 1986).

If different items for a scale have different ranges of values, they should be standardized before combining so that no one item dominates the scale by having a wider range of values. For example, a socioeconomic status scale that combined measures of income (ranging from 0 to millions of dollars) with measures of educational attainment (ranging from 0 to 20 years of schooling), any contribution of educational attainment would be overwhelmed by variation in income (Treiman 2009). See chapter 5 for more on standardized scores, and "Distributions of Scales or Indexes" below for an example of incorrect scale creation from variables with very different ranges of values.

Some scales are constructed by assigning different weights to different patterns of responses, reflecting the idea that some suggest a weak degree of the construct while other response patterns reflect stronger degrees of that same construct. For example, Mooney and Lee (1995) developed a scale of permissiveness of states' abortion regulations by assigning a lower weight (permissiveness score) to states that allow abortion only "when the pregnancy results from rape or incest" than for states that allow it for any reason, which encompasses the more stringent criteria. See Babbie (2006) or Treiman (2009) for more on constructing scales and indexes.

Before you analyze and write about numbers, familiarize yourself with how the data were collected, to ensure that you make sensible choices about calculations, creation of new variables, consistency checks, and ways of presenting the information. See Miller (2013b) for an in-depth discussion and structured set of exercises for to getting to know each of the variables you will use from your data set.

SPECIFYING UNITS

To interpret your numbers and facilitate comparison with those from other sources, specify the units of analysis and the units and systems of measurement. Make a note of the units for any numeric facts you collect from other sources so you can use and interpret them correctly.

Dimensions of Units

UNIT OF ANALYSIS

The unit of analysis (also known as the "level of aggregation") identifies the level at which numbers are reported. If you measure poverty in number of persons with income below some threshold, your unit of analysis is a person. If you count families with income below a threshold, your unit of analysis is a family. Poverty can also be calculated at the census tract, county, state, or national level. Each approach is valid, but values with different levels of aggregation cannot be compared with one another. In 2012, there were 46.4 million poor *people* but 9.5 million poor *families* in the United States (DeNavas-Walt, Proctor, and Smith 2013). When collecting information to compare against your data, look for a consistent level of aggregation.

Level of aggregation pertains to most quantitative topics. For instance, cancer penetration can be measured by the number of cells affected within a particular organ, cancerous organs in a patient, people afflicted within a family or town, or the number of towns with cases. Avoid confusion by stating the level of aggregation along with the numbers:

"Mr. Jones's cancer had metastasized, affecting three major organ systems."

"Breast cancer is widespread in the area, with at least five towns having prevalence rates well above the national average."

UNITS OF MEASUREMENT

There are two aspects of units of measurement: scale and system of measurement. Both are critical for your results to be interpretable.

Scale

Scale, or order of magnitude, refers to multiples of units. Are you reporting number of people, number of thousands of people, millions of people, or some other scale of measurement? Consider the following error of omission, which had data analysts scratching their heads: a utility company observed a sudden and substantial drop-off in demand for electricity at the end of 1978 between periods of fairly steady demand (figure 4.1). At first, researchers looked for a real, causal explanation. The region had several major industrial clients. Had one of them closed? Had a competing electricity provider opened up shop? In fact, the apparent drop was due to a change in the scale of units used to report electricity demand, from hundreds of megawatt-hours to thousands of megawatt-hours. The actual amount of electricity used was fairly stable.

System of measurement

There is plenty of room for confusion in a world where metric, British, and other systems of measurement coexist. Virtually every dimension of our experience that is quantified—distance (feet or meters), mass (pounds or kilograms), velocity (miles per hour or kilometers per hour), volume (quarts or liters), money (dollars or Euros or pesos or yen), dates (Christian or Jewish or other calendar), and time (standard time or daylight saving time or military time)—can be measured using any of several systems.

The embarrassing experience of the Mars Climate Orbiter in 1999 is a classic illustration of what can happen if system of measurement is not specified. Engineers who built the spacecraft specified the

Electricity demand, Industrytown, 1978–1979

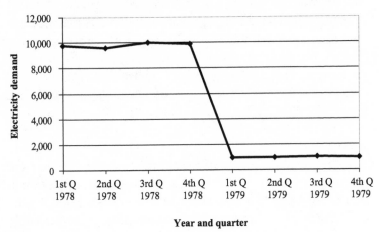

Year and quarter

Figure 4.1. Effect of a change in scale of reported data on apparent trends

spacecraft's thrust in pounds, which are British units. NASA scientists thought the information was in newtons, which are metric units. The miscalculation went overlooked through years of design, building, and launch, and the spacecraft missed its target by roughly 60 miles (Pollack 1999). Even rocket scientists make basic, easily avoidable mistakes about units. Don't emulate them.

Writing about Units

Incorporate units of observation and measurement into the sentence with the pertinent numbers.

> "In 2012, annual per capita income in the United States was $28,051 (US Census Bureau 2014)."
> *This sentence includes information on units of measurement (dollars), units of observation (per capita means "for each person"), and the W's (what, when, where).*

Definitions of Units

Familiar units such as dollars, numbers of people, and degrees Fahrenheit can be used without defining them first. Likewise, if you are writing for experts in your field, you can usually skip an explanation of commonly used units in that field. However, remember that what is familiar to a group of experts might be completely Greek to people from other disciplines or to an applied audience; adjust your explanations accordingly. For instance, measures such as constant dollars, age-standardized death rates, or seasonally adjusted rates will need to be defined for people who do not work with them routinely. Regardless of audience, provide a standard citation to explain the calculation.

Inexperienced writers often fail to explain their units precisely or completely, particularly for common but tricky measures that express relationships between two quantities such as parts of a whole, the relative size of two groups, or rates. For variables that are measured in percentages, proportions, rates, or other types of ratios, include phrases such as "of ___," "per ___," or "compared to ___" so the values can be interpreted. See chapter 5 for more explanations and examples of how to express units in your writing.

Ratios

Ratios are simply one number divided by another. They measure the relative size of two quantities. Sometimes the quantities in the numerator and denominator are mutually exclusive subsets of the same whole. For instance, the sex ratio is defined as number of males per 100 females. Some ratios divide unrelated quantities. For example, population density is number of people per land area (e.g., per square mile).

Proportions, percentages, and fractions

Proportions (and their cousins, percentages) are a special type of ratio in which a subset is divided by the whole, such as males divided by total population. Percentages are simply proportions multiplied by 100. If the proportion male is 0.50, then 50% of the population is male. Do not make the all-too-common mistake of labeling proportions and percentages interchangeably; they differ by a factor of 100.

For measures such as proportions, percentages, and fractions that compare parts to a whole, make it clear what "the whole" comprises. In technical terms, what does the denominator include? If I had a nickel for every time I have written "percentage of what?" in the mar-

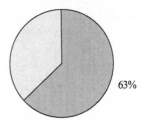

**a. Voter participation,
1996 US presidential election
% of voting age population**

47%

**b. Voter participation,
1996 US presidential election
% of registered voters**

63%

Figure 4.2. Effects of different referent group definitions on percentage calculations
Source: International Institute for Democracy and Electoral Assistance 1999

gins of a paper, I would be rich. Consider the following statistics on voter turnout from figure 4.2:

> *Poor:* "In 1996, the voter turnout for the presidential election was 63%."
> *Is voter turnout measured as a percentage of the voting age population (figure 4.2a), or as a percentage of all registered voters (figure 4.2b)? Because some people of voting age are not registered to vote, the two measures cannot be directly compared, but to avoid that mistake, readers need to know which measure is reported.*
> *Better:* "In 1996, 63% of the 146 million registered voters participated in the US presidential election."
> *Here, number of registered voters is identified as the basis of comparison, so this number can be compared with equivalently defined voter participation rates for other years or places.*
> *Best:* "In 1996, 63% of the 146 million registered voters participated in the US presidential election. As a percentage of the voting age population (197 million people), however, voter turnout was only 47%, revealing a large pool of potential voters who did not participate."
> *By specifying the denominator for each statistic, this version gives the information needed to assess comparability of numbers from other sources. The description is enhanced by contrasting*

the two measures of voter participation and explaining how they differ conceptually.

When relationships involve two or more variables, such as the association between poverty and age group, "percentage of what?" is more complicated. In cross-tabulations like table 4.1, there are several possible definitions of the whole against which some part is being compared. Depending on the question you are asking, you might report the percentage of the *total population* that is in each of the six possible poverty/age group combinations (table 4.1a), the percentage of *each poverty category* that falls into each age group (table 4.1b), or the percentage of *each age group* that is poor (table 4.1c).

Just over one-third of the poor are children (<18 years old; table 4.1b), but more than one-fifth of children are poor (table 4.1c).[4] These values have completely different meanings, so specify what subgroup is being compared to what whole entity.

Poor: "In 2012, there were a lot of poor children in the United States (21.8%)."
Does 21.8% refer to poor children out of all people, poor children out of all poor people (both incorrect in this case), or poor children out of all children (correct)?
Better: "In 2012, nearly 22% of children in the United States were poor."
The referent group (children) for the percentage is stated, clarifying interpretation of the statistic.
Best: "In 2012, nearly 22% of children were poor, compared to about 14% of people aged 18 to 64 years, and 9% of those aged 65 or older."
This sentence makes it clear that the observed poverty rate among children is higher than that for other age groups. Both the units of analysis (persons) and measurement (percentage of the age group) are specified and are consistent with one another, hence the comparison is correctly done.

Rates

Rates are a type of ratio that calculates the speed or frequency with which an event or circumstance occurs per unit of time or population.

Table 4.1. Three tables based on the same cross-tabulation: (a) Joint distribution, (b) Composition within subgroups, (c) Rates of occurrence within subgroups

a. Poverty by age group, United States, 2012

Age group (years)	Thousands of persons (% of total population)		
	Poor	Non-poor	Total
<18	16,073 (5.2%)	57,646 (18.6%)	73,719 (23.7%)
18–64	26,497 (8.5%)	167,145 (53.8%)	193,642 (62.3%)
65+	3,926 (1.3%)	39,361 (12.7%)	43,287 (13.9%)
Total	46,496 (15.0%)	264,152 (85.0%)	310,648 (100.0%)

b. Age distribution (%) by poverty status, United States, 2012

Age group (years)	Poor		Non-poor		Total	
	Number (1,000s)	% of all poor	Number (1,000s)	% of all non-poor	Number (1,000s)	% of total pop.
<18	16,073	34.6%	57,646	21.8%	73,719	23.7%
18–64	26,497	57.0%	167,145	63.3%	193,642	62.3%
65+	3,926	8.4%	39,361	14.9%	43,287	13.9%
Total	46,496	100.0%	264,152	100.0%	310,648	100.0%

c. Poverty rates (%) by age group, United States, 2012

Age group (years)	# Poor (1,000s)	Total pop. (1,000s)	% poor within age group
<18	16,073	73,719	21.8%
18–64	26,497	193,642	13.7%
65+	3,926	43,287	9.1%
Total	46,496	310,648	15.0%

Source: DeNavas-Walt, Proctor, and Smith 2013.

For example, velocity is often measured in miles per hour (distance divided by time), death rates as number of deaths per 100,000 people within a specified period of time (Lilienfeld and Stolley 1994). Report the units and concepts in both the numerator and denominator, and the time interval if pertinent.

A common error is to confuse a death rate for a particular subgroup with deaths in that group as a proportion of all (total) deaths, as in the following example.

Poor: "In the United States in 2010, almost one-third died from cardiovascular disease (American Heart Association 2014)."
One-third of what? Deaths (correct, in this case)? People? By failing to specify one-third of what, the author leaves this sentence open to misinterpretation: the fraction of all deaths that occurred due to cardiovascular disease is not the same as the death rate (per population) from cardiovascular disease.

Poor (version 2): "About one-third of people died from cardiovascular disease in the United States in 2010 (American Heart Association 2014)."
This version is even worse because it implies that one out of every three living people died of cardiovascular disease in 2010—a figure that would translate into roughly 99 million cardiovascular disease deaths in the United States that year. In fact, the actual death toll from all causes combined was only 2.46 million. Don't make the mistake of thinking that "one-third is one-third," without specifying the "of what?" part of the fraction.

Best: "In 2010, about 725,000 people died of cardiovascular disease (CVD) in the United States, accounting for roughly one out of every three US deaths that year. The death rate from CVD disease was 235.5 deaths per 100,000 people (American Heart Association 2014)."
This version conveys that cardiovascular disease accounted for approximately one-third of deaths that year. Mentioning the death rate clarifies the other perspective on cardiovascular disease—the annual risk of dying from that cause.

EXAMINING THE DISTRIBUTION OF YOUR VARIABLES

As you write about numbers, you will use a variety of examples or contrasts. Depending on the point you want to make, you may need

- a typical value, such as the average height or math score in a sample;
- an expected change or contrast, such as a proposed increase in the minimum wage;
- the extremes; for example, the maximum possible change in test scores.

To tell whether a given value is typical or atypical, or a change is large or modest, you need to see how it fits in the distribution of values for that variable *in your data and research context*. For instance, the range of math scores will be smaller (and the average higher) in a group of "gifted and talented" students than among all students in an entire school district; which you would use depends on your research question and the available data.

Avoid treating the variables in your analyses as generic, instead taking the time to become familiar with the specific concepts and units for each of your variables, which will help you identify reasonable levels and ranges of values for each variable in your analysis. The range of credible values for a particular variable can be affected by definitional limits, what is conceptually plausible, and the context of measurement. For instance, the percentage of a whole by definition must fall between 0 and 100, but a percentage *change* can be negative or exceed 100 (chapter 5). Many other variables also cannot assume negative values. For example, an index constructed by summing 10 items each of which could range from 0 to 2 will have a mathematically-defined minimum of 0 and maximum of 20.

The conceptually plausible range is topic-specific, as with infant birth weight, which is limited by anatomical constraints. It is also unit-specific: for example, live births in the United States have birth weight in *grams* that range from about 400 to 5,900 (table 6.4), but the corresponding range in *ounces* is from 14 (less than 1 pound) to 208 (13 pounds). Finally, context (when, where, and who is in a sample) will affect the range of reasonable values, e.g., current income in the United States versus the United States in 1900 or Vietnam today.

Do not simply run statistics on your data without stopping to learn their substantive, real-world meaning or to check the range of values (Miller 2013b). Before you select values for calculations or as case examples, acquaint yourself with the literature on your topic and context so you are aware of what values make sense for your variables. Then use exploratory data techniques such as frequency distributions, graphs, and simple descriptive statistics to familiarize yourself with the distributions of the variables in your data.

Level

Start by checking the level of each variable, taking into account the topic, units, and context. Consider the following examples of how not all variables can take on all numeric values: A value of 10,000 makes sense in at least some contexts (places, times, or groups) for annual family income in dollars, the population of a census tract, or an annual death rate per 100,000 persons. However, with rare exceptions, a value of 10,000 does *not* make sense for hourly income in dollars or birth weight in grams, and never fits number of persons in a family, a Likert-type item, a proportion, or an annual death rate per 1,000 persons. A value of −1 makes sense for temperature in degrees Fahrenheit or Celsius, *change* in rating on a 5-point scale, change in a death rate, or percentage change in income, but is completely nonsensical for temperature in degrees Kelvin, number of persons in a family, a proportion, or a death rate.

This story about the experience of a young research trainee will illustrate the importance of this simple step in learning about your data: She came to me in the ninth week of a ten-week training program, puzzled by the results of her statistical analysis. She was analyzing predictors of birth weight using a nationally representative survey sample from a developing country circa 2002, which she had downloaded from a research data website but hadn't cleaned or evaluated before analyzing it. In the sample, values of the birth weight variable ranged up to 9,999 with a mean of 8,000. Had she taken the time to look up the expected range of values for that concept (birth weight) and units (grams), she would have immediately seen a red flag because 9,999 grams is roughly 22 pounds, which is a typical weight for a one-year-old, not a newborn!

Missing Values

Next, acquaint yourself with the missing value codes for each of your variables before evaluating their distributions for plausibility. Again, consider the story of the research trainee mentioned earlier: A second warning sign was that two-thirds of her sample had a birth weight value of 9,999—a very high value for such a substantial share of a sample, and one that is unlikely to be explained by either outliers or data entry errors alone. By examining the study documentation and questionnaire, she discovered that this distribution occurred due to a skip pattern designed to minimize recall bias, such that only mothers of children under age 5 years were asked about birth weight; see "Missing by Design" in chapter 10 for more on skip patterns. Children aged 5 through 17 years should have been omitted from her analytic sample because the outcome variable was missing for such cases, and were thus assigned the missing value code of 9999 in the data set she had downloaded. After discovering these issues, she was forced to rerun all of her statistical analyses at the last minute to complete her final paper. Avoid such problems by familiarizing yourself with how the data on your topic were collected so you can identify any missing value codes and treat them as such in the electronic database before examining the distribution of values for which there were valid (nonmissing) responses.

Minimum, Maximum, and Range

After identifying missing values and confirming that the values of your variables make sense, examine the minimum and maximum observed values and the range—the difference between them. Assess the *actual* range of values taken on by your variables, not just the theoretically possible range. For instance, as noted above, the CESD is composed of 20 questions about frequency of certain depression symptoms, each scaled from 0 to 3. Although in theory the CESD scale could range from 0 to 60, in the general population the mean is between 8 and 10, and scores above 20 are rarely observed (Radloff and Locke 1986). Thus using a change of 25 points as an illustrative example would be unrealistic. See chapter 8 for more discussion of out-of-range values and contrasts.

Measures of Central Tendency

The second aspect of distribution to consider is central tendency—the mean, median, and/or modal values, depending on the type of variable in question. The mean is usually the arithmetic average of the values under study, calculated by adding together all values and dividing that sum by the number of cases.[5] The median is the middle value (at the 50th percentile) when all values are ranked in order from lowest to highest. The mode is the most common value—the value observed the most frequently of all values in the sample. Any of the measures of central tendency can be used for continuous variables, but neither the mean nor the median makes sense for categorical (nominal or ordinal) variables.

Although the mean is the most widely used of the three measures of central tendency for continuous variables, before you rely on it to characterize your data, observe the distribution of values. The mean does not always represent the distribution well. In figures 4.3a, b, and c, the mean value (6.0) would be an appropriate example, although in figure 4.3c it is no more typical than any of the other observed values. (See "Variability" below for an explanation of "SD.")

In figure 4.3d, however, the mean (still 6.0) is not observed for any cases in the sample and hence is not a representative value. Likewise, in figure 4.3e, the mean (again 6.0) is atypical. If such a pattern characterizes your data, the mode would be a better choice of a typical value. Another caution: the mean can be biased if your sample has one or two outliers—values that are much higher or much lower than those in the rest of the sample. Use Tukey's box-and-whisker or stem-and-leaf techniques to display deviation from the mean or identify the presence of outliers (see chapter 7 or Hoaglin et al. 2000). See chapter 10 for how to describe treatment of outliers.

Variability

Another important consideration is the variability in your data, the extent to which values are spread around the mean. It is usually summarized by the variance or standard deviation (SD). For example, a difference of 2 points is "bigger" or more meaningful in a distribution that is tightly clustered around the mean (e.g., SD = 1.07; figure 4.3a)

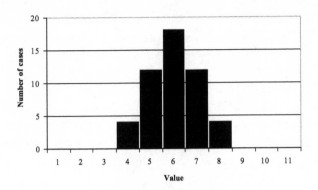

a. Normal distribution
(Mean = 6.0; SD = 1.07)

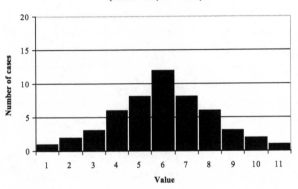

b. Normal distribution
(Mean = 6.0; SD = 2.18)

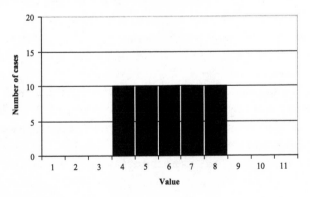

c. Uniform distribution
(Mean = 6.0; SD = 1.43)

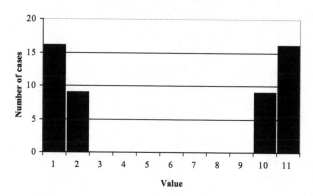

d. Polarized bimodal distribution
(Mean = 6.0; SD = 4.71)

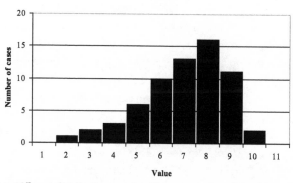

e. Skewed distribution
(Mean = 6.0; SD = 1.76)

Figure 4.3. Different distributions, each with a mean value of 6.0. (a) Normal distribution; (b) Normal distribution, higher SD; (c) Uniform distribution; (d) Polarized bimodal distribution; (e) Skewed distribution

than in one with more spread (e.g., SD = 2.18; figure 4.3b). In the first instance, 2 points is 1.86 standard deviations from the mean, placing it farther out in the distribution than in the second instance, where 2 points is less than one standard deviation. To convey the position of a given value in the overall distribution, report its rank, percentile, or z-score (chapter 5).

Alternatively, use the "five number summary" (Moore 1997)—the minimum value, first quartile value (Q1), median, third quartile (Q3), and maximum value—to illustrate the spread. The minimum and maximum values encompass the range, while the interquartile range (Q1 to Q3) shows how the middle half of all values are distributed within the full span of the data. These numbers can be reported in tabular form or a box-and-whisker chart such as figure 7.10.

Distributions of Scales or Indexes

Remember to examine the distribution of scales, indexes, or other variables that were created (whether by you or others) by combining or transforming variables. Consider the example of an acculturation scale that was created from measures of time spent in the United States, whether the respondents were first, second, or third generation immigrants, and what language they spoke at home. A histogram (figure 4.4) revealed that the distribution of that scale for a nationally representative sample of Latinos was highly unusual, with three small approximately-normal distributions just above values of 0.0 points, 2.0 points, and 4.0 points, gaps between those distributions, and spikes at exact values of 6.0 and 7.0.

Closer examination revealed that the scale had combined one continuous variable that was measured as a proportion (bounded between 0 and 1) with two categorical variables that each took on integer values of 0 and 2, and another categorical variable that assumed values of 0, 1, and 2. In the flawed acculturation scale, the variables with 2-unit spans overwhelmed any variation in the variable measured as a proportion. When constructing scales, either combine variables that share a common level of measurement and coding scheme or standardize the variables before combining them so that the scale is not dominated by variable(s) with higher values or wider variances; see "Indexes and Scales" above.

In addition, the component acculturation variables were related to one another such that it was mathematically impossible for anyone to have values between 1.0 and 2.0, between 3.0 and 4.0, between 5.0 and 6.0, or between 5.0 and 6.0, yielding the bizarre distribution shown in figure 4.4. For all of those reasons, the variables should have been recognized as tapping distinct dimensions of acculturation and analyzed separately rather than being combined into a scale—a set

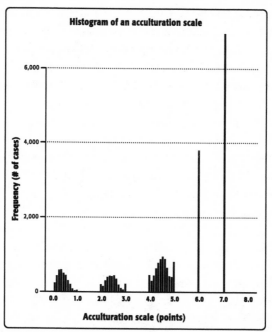

Figure 4.4. Histogram of a scale constructed from mixed continuous and categorical variables

of issues that became evident when the distribution of the scale was graphed.

Transformed Variables

In some situations, you may decide to transform some of your variables, using multiples of the original units or taking logarithms, for example. If so, examine the distribution of the transformed variable before you select examples and contrasts.

DEFINING STANDARD CUTOFFS AND PATTERNS

Comparison against a standard is a useful way of assessing whether a particular numeric value of a variable is commonplace or exceptional, or an observed pattern typical or unusual for the topic under study.

What Is a Standard?

Standards include cutoffs, average patterns, and records that define the highest and lowest observed values. Some cutoffs are based on physical principles: the properties of water change below 32°F. Other cutoffs are based on social or legal convention: 21 years as the minimum drinking age in most US states; a specified dollar amount as the Federal Poverty Level for a family of certain size and age composition, adjusted annually for inflation. Commonly used standard patterns include the J-shaped age pattern of mortality, the seasonal pattern of employment for a given occupation, and growth charts for height and weight of children from birth to adulthood.

Some standard patterns are empirically derived, like the mean temperature for New York City on January 1 calculated from several decades of data, or the median height for 12-month-old girls computed from a national sample. Other standards are conventions agreed upon by experts in the field. The Bureau of Labor Statistics calculates the Consumer Price Index (CPI) with 1982–1984 as the base period when CPI = 100 (US Bureau of Labor Statistics 2007). At the start of the twenty-first century, demographers and epidemiologists at the Census Bureau and Centers for Disease Control and Prevention replaced the 1940 age structure with that for 2000 as the basis for their "standard million" for all age adjustments (Anderson and Rosenberg 1998; CDC 1999).

Often the initial choice of a standard is somewhat arbitrary—it doesn't really matter which year or population is chosen. However, the same convention must be used to generate all numbers to be compared because the results of an age adjustment or a constant dollar calculation vary depending on which standard was used. To ensure comparability, read for and specify the context (W's), units, and methods used to apply the standard.

Although the term "standard" implies uniformity of definition, there is a striking variety and malleability of standards. As of January 1, 2002, the National Weather Service began using the average conditions from 1971 through 2000 to assess temperatures, replacing the relatively cool decade of the 1960s with the relatively warm decade of the 1990s. Thus a temperature that would have been interpreted as unusually warm if it had occurred on December 31, 2001, when the old standard was in use might have been considered normal or even cool if it had occurred one day later, when the new standard was in place. Periodically, the standard referent year for constant dollars, the standard population for age adjustment of death rates, and human growth standards are also updated.

Standards also vary by location. What constitutes a normal daytime high temperature in Los Angeles in January is quite different from the normal daytime high for that month in Chicago or Sydney. A different age pattern of mortality is observed in the United States and other developed countries today (where chronic diseases account for most deaths) than in Afghanistan and other less developed countries (where infectious and accidental causes of death predominate; Omran 1971; Preston 1976). In addition, standards or thresholds may differ according to other characteristics. For instance, Federal Poverty Levels ("poverty thresholds") vary by family size and age composition: the threshold for a single elderly person is lower than for a family with two adults and two children.

Why Use Standards?

Standards are used to assess a given value or pattern of values by comparing against information about pertinent physical or social phenomena. Such comparisons factor out some underlying pattern to help ascertain whether a given value is high, low, or average.

- Does a particular value exceed or fall below some cutoff that has important substantive meaning? For instance, alcohol consumption is illegal only below a certain age.
- Is an observed trend consistent with other concurrent changes? Did college tuition rise faster than the general rate of inflation, for example?
- Is the pattern for a given case typical or unusual? For instance, is a child growing at the expected rate? Is he significantly above or below the average size for his age? If so, is that gap increasing, decreasing, or stable as he gets older?
- Is a particular change expected based on cyclical patterns or due to some other factor? Did a new law reduce traffic

fatalities, or is such a decline expected at the time of the year the law took effect, for example?

In addition to some measure of average value, standards often include information on range or other aspects of distribution. Weather standards mention record high and low temperatures as well as averages. Growth charts for children typically show the 10th, 25th, 50th, 75th, and 90th percentiles for age and sex.

Selecting an Appropriate Standard

For many nontechnical topics, cutoffs and patterns are part of cultural literacy: most Americans grow up knowing they'll have the right to vote when they turn 18 and can retire at 65. The freezing point of water, the concept of 50% or more as a democratic majority, and that it is warmer in summer than winter are also generally known. For more technical issues or if precise values are needed, read the related literature to become familiar with standard cutoffs, patterns, or standardization calculations for your topic. See chapter 5 for additional discussion of how to use standards in numeric contrasts.

Where and How to Discuss Standards

For a lay audience, use of standards is a behind-the-scenes activity. Report the conclusions but not the process. Do children from low-income families grow normally? Did this year's employment follow the typical cyclical pattern? Because standards vary, mention the date, place, and other attributes of the standard you are using: "in constant 2000 dollars," "compared to the normal daily high for November in Chicago," or "below the 2012 poverty threshold for a family of two adults and one child."

For a scientific audience, specify which standard has been applied. Explain cutoffs, standard patterns, or standardization processes that are unfamiliar to your audience, and consider using diagrams (see "Reference Lines" in chapter 7) or analogies (chapter 8) to illustrate. If you have used a cutoff to define a categorical variable, explain that in your data and methods section. If the approach is new or unusual, include the formula and a description in the methods section, a footnote, or an appendix. For lengthy calculations or explanations, cite a published source where readers can find the details.

CHOOSING A FITTING NUMBER OF
DIGITS AND DECIMAL PLACES

How many digits and decimal places should you include as you write about numbers? Too many are cumbersome and uninformative and may overstate the precision with which the data were originally measured. Too few may not meet your readers' objectives.

Precision of Measurement and Number of Significant Digits

Established scientific guidelines specify how the precision of measurement limits the appropriate number of digits to report for both measured values (raw data) and calculations (Logan 1995; NIST 2000) — a concept referred to as "significant digits" or "significant figures."[6] If a ruler is marked with increments for each millimeter (tenth of a centimeter or 0.1 cm), one can "eyeball" to approximate halfway between those marks, or to the nearest 0.05 cm. All values measured with that ruler and all calculations based on those data should be reported with no more than two decimal places. Likewise, if you are averaging income that was originally collected in multiples of thousands of dollars, you can report one level down (e.g., hundreds of dollars), but not single dollars and cents.

It is surprising how often calculators and computers seem to magically enhance the implied level of detail and precision of numeric information. Just because your calculator or computer output displays eight or more digits or decimal places does not mean that information is accurate. Make sure the number of decimal places in reported calculations is consistent with the precision of the original measurement. See Logan (1995) for a comprehensive discussion of how measurement and calculations determine the appropriate number of significant digits.

The number of significant digits is often fewer than the total number of digits displayed because of the presence of leading or trailing zeroes. Leading zeroes are those found to the right of the decimal point, before the first "significant" (nonzero) numeral, e.g., the five zeroes between the decimal point and the 2 in the numeral "0.0000023." They serve as placeholders that convey the scale of the number, in this case, millionths. Likewise trailing zeroes are placeholders for thousands, millions, or billions. For example, the six zeroes used to convey the scale in "285,000,000" rarely reflect an exact count down to the single unit. Eliminate them by rounding to that scale—just write "285 million."

Number of Digits

Guidelines about significant digits set an upper limit on the appropriate number of digits and decimal places for your data. Rarely, however, are all those digits needed to make your point. For most purposes, "approximately six billion persons" is an adequate description of the world's population at the turn of the millennium. The number "6,049,689,552 persons" is more detail than all but the most persnickety demographer could want to know, and almost certainly exaggerates the precision of the estimate. To provide a bit more detail, write "6.0 billion" or "6.05 billion" persons. Information about the last few hundred persons is not very informative when the overall scale is in the billions. Often, the main point is the similarity of two or more numbers; if so, don't display extra digits merely to reveal tiny differences. And you don't want your readers pausing to count commas to figure out the scale of the number: "Let's see. Three commas . . . that's billions."

In most prose, two to four digits are enough to illustrate your point without too much detail. Consider these three ways of reporting results of a very close election:

> *Poor:* "In the recent election, Candidate A received 2,333,201 votes while Candidate B received 2,333,422 votes."
> *To figure out who won and by how much, readers must wade through a lot of digits and then do the math themselves.*
> *Better:* "Candidate B won the recent election over Candidate A, with 2,333,422 votes and 2,333,201 votes, respectively."
> *This version conveys who won, but readers must still calculate the margin of victory.*

Best: "In an extremely close election, Candidate B eked out a victory over Candidate A, receiving only 221 more votes out of more than 4.6 million votes tallied—a margin of less than 100th of 1% of the total votes."

By interpreting the results of the calculations, the key point is far more accessible than in the previous two versions. Who won and by how much? Phrases such as "extremely close" and "eked" communicate how narrow the victory was much more effectively than just reporting the numbers, and none of the numbers presented include more than three digits. If the exact number of votes is needed, accompany the description with a simple table reporting the tallies.

Number of Decimal Places

In general, include the smallest number of decimal places that suit the scale of your numbers—in many cases, none. If most of the variation is in the thousands, millions, or billions, detail in the second decimal place won't add much to the main point: Do we really need to know the federal budget down to the penny? Or believe that it is measured with that level of precision? On the other hand, for numbers less than 1.0, a couple of decimal places are needed for the variation to be visible: because of rounding, differences between tax assessment rates of 0.072% and 0.068% won't be evident unless you show three decimal places. And p-values or test statistics can't be evaluated if they are rounded to the nearest whole number.

Having chosen an appropriately modest number of digits and decimal places for the text, do not turn around and report a zillion digits in your accompanying tables. Yes, tables are good for detail, but there is a limit to what is useful. Don't overwhelm your readers with a table of six columns and 20 rows, each cell of which contains eight digits and six decimal places, particularly if your associated narrative discusses each of those numbers rounded to the nearest two digits.

Use the recommendations in table 4.2 to decide what level of detail is suitable for the types of numbers you report, then design your tables and prose accordingly. On occasion, you might exceed the recommended number of digits or decimal places, but do so only if

Table 4.2. Guidelines on number of digits and decimal places for text, charts, and tables, by type of statistic

Type of Statistic[b]	Text or Chart			Table		
	Total digits[a]	Decimal places	Examples	Total digits	Decimal places	Examples
Integer[b]	3 to 4	Not applicable	7 million 388	Up to 6	Not applicable	7,123 thousand[c] 388
Rational number	3 to 4	1 to 2	32.1 -0.71	Up to 6	Up to 4; enough to show 2 significant digits	32.1 -0.71 0.0043
Percentage	3 to 4	1 if several #s would round to same value; otherwise none	72% 6.1%	3 to 4	2 if several #s would round to same value; otherwise 1	72.1% 6.12%
Proportion	Up to 3	3 if several #s <0.10; otherwise 2	.36 .0024	Up to 3	3 if several #s <0.10; otherwise 2	.36 0.024

(continued)

Table 4.2. (continued)

Type of Statistic	Text or Chart			Table		
	Total digits[a]	Decimal places	Examples	Total digits	Decimal places	Examples
Monetary value	3 to 4	None for large denominations; 2 for small	$5 million $12.34	3 to 4	None for large denominations; 2 for small	$5 million $12.34
Ratio	Up to 3	1 to 2	12.7 0.83	Up to 4	2 if one or more ratios <1.0; otherwise 1	12.71 0.83
Test statistic	3 to 4	2	$\chi^2 = 12.19$ $t = 1.78$	3 to 4	2	$\chi^2 = 12.19$ $t = 1.78$
p-value	Up to 3	2	$p = 0.06$ $p < 0.01$	Up to 3	2 for values ≥ 0.01; 3 for values < 0.01	$p = 0.06$ $p < 0.001$

Note: See Logan (1995) for considerations on appropriate number of digits for calculations.

[a] Including decimal places. If number of digits exceeds this value, round the number or change scale.

[b] Integers include the positive and negative counting numbers and zero (Kornegay 1999). By definition, they have no decimal places.

[c] The word "thousand" (or other unit name) would appear only in the column head, not in the table cells; see table 4.1 for an example.

fewer will not accomplish your purpose. For some types of numbers, there are well-established conventions about how many decimal places to include:

- Monetary denominations include two decimal places to show value to the nearest cent, except for values of $1 million or greater, when decimal places are usually superfluous. Even on my bank statement (with a balance far below a million dollars), I rarely look at how many cents I have.
- Proportions include three decimal places if several values presented are less than 0.10; otherwise, two will suffice.
- Percentages often don't need any decimal places unless the values are very similar to one another (e.g., 5.9 and 6.1, or 67.79 and 67.83). In such cases, include one or two decimal places, as shown.
- Test statistics (e.g., t-statistic, X^2, or F-statistic) require two decimal places to compare them against critical values.
- p-values conventionally include two decimal places, although three may be shown if $p < 0.01$, to display one significant figure.

In general, I recommend choosing the numbers of digits and decimal places depending on the scale of the numbers and precision with which they are measured, aiming for no more than four total digits for each value reported in prose, and no more than six (ideally fewer) in tables. Possible exceptions include tables of raw data in reports intended as standard data sources for public use, and "dazzle" statistics to catch your audience's attention with a single showy number. In the latter case, you might report the federal budget down to the last cent, then remind them that such detail may be fun but is probably unwarranted. Report fewer digits for general background statistics than for detailed quantitative analyses, and use a consistent scale and number of decimal places within a series of numbers to be compared.

Changing Scale, Rounding, and Scientific Notation

One way to reduce the number of digits you report is to change the scale of the numbers, rounding to the nearest million or thousandth, for example. To decide on an appropriate scale, consider the highest and lowest values you need to report, then choose a scale that gracefully captures most values using three to four numerals. In 2010, the populations of the 50 United States ranged from approximately 564 thousand people in Wyoming to 37.3 million people in California (US Census Bureau 2010). Because most states' populations exceed one million, round the figures to the nearest million with one or two decimal places to accommodate both small and large states' populations: 0.6 million people in Wyoming, 37.3 million in California.

Changing scale also can help reduce the number of digits for proportions or other statistics that have several leading zeroes without losing any meaningful information. Convert proportions to percentages to save two decimal places, sparing readers the chore of counting zeroes to assess scale. For instance, a proportion of 0.0007 becomes 0.07% or could be rounded to 0.1%. This averts the common mistake of calling proportions percentages, and makes the scale of the numbers easier to grasp. Or, for a chemical sample weighing 0.0000023 grams, report it as 2.3 micrograms; in a nonscientific piece, include a note that one microgram equals one millionth of a gram. In a lab report or other document for a biological or physical science audience, use scientific notation—another convention for succinctly presenting only the meaningful digits of a number; write 2.3×10^{-6} grams.

Numbers versus Numerals

A few more technical points: some editors require that numbers under 10 and units (e.g., "percent," "million") be spelled out rather than reported in numeral form. These guidelines vary across disciplines, so consult a manual of style for your field. Spell out numbers at the beginning of a sentence: "Thirty percent of all deaths were from heart disease," rather than "30 percent of all deaths were from heart disease." Or rephrase the sentence to put the number later in the sentence. "Heart disease accounted for 30% of all deaths." Whenever possible, separate distinct numeric values with more than a comma, using symbols or names for units (e.g., %, grams) or adverbs (e.g., "approximately," "nearly") to help readers distinguish where one number ends and the next begins. For instance, replace "100, 320, and

799 grams, respectively" with "100 grams, 320 grams, and 799 grams, respectively." See University of Chicago Press (2010) or Alred et al. (2000) for additional technical writing guidelines.

CHECKLIST FOR FIVE MORE BASIC PRINCIPLES

- Familiarize yourself with each of your variables.
    Are they categorical or continuous?
    - If categorical, are they nominal or ordinal?
    - If continuous, are they ratio or interval?
    Are they single- or multiple-response?
    Are they scales or indexes, and if so, how are they calculated from the source variables?
- Know the units of measurement for each variable.
    Check units:
    - Level of aggregation or unit of analysis.
    - Scale or order of magnitude.
    - System of measurement, such as British, metric, or other.
    - Composite, standardized, or transformed variables.
    Check comparability of units within your work and with that of others.
- Use descriptive statistics and graphs to examine the distribution of your variables: range, central tendency, variance, and symmetry in order to identify typical and atypical values or contrasts.
- Consider standard cutoffs, distributions, or historic records.
    Find out which are used for your topic.
    Cite references for standards.
    Put details in appendixes or footnotes.
- Pick an appropriate number of digits and decimal places, taking into account
    Precision with which the data were originally measured.
    Objectives of your work.
    - Aim for four digits with up to two decimal places for each numeric value in the text and charts, one to two more in tables (see table 4.2 for guidelines).
    - Round or change the scale to reduce number of digits or leading or trailing zeroes.

Part II

TOOLS

In this part, I introduce some basic tools for calculating and comparing numbers, designing tables and charts, and choosing numeric examples and analogies. Used in concert with the principles described in the previous few chapters, these tools will help you develop effective ways to present quantitative information.

To explain the purpose and application of the tools, I have written these chapters in a "teaching" style. In general, this is *not* how you will write about numbers. For example, your readers don't need a detailed description of how you approached writing up $p = 0.08$, the steps to calculate percentage change, that you right-justified the numbers in your table, or why you chose a stacked bar chart rather than a pie chart. Make those decisions, do those calculations, and create your charts or tables to function well, but don't write about how or why you did so. Instead, present the fruits of those labors, following the examples and guidelines throughout this book.

An important exception is when you are writing about numbers in a problem set or paper for a research methods or statistics course. In those instances you might be asked to show your calculations and explain your thought process to demonstrate that you have mastered the corresponding concepts and skills. The methods section of a scientific paper also often includes a description of analytic strategy; see chapter 10. Check with your professor to find out whether to include this information in your course assignments. If you later revise a course paper for publication or presentation to another audience, remove most of the "teaching statistics" material and focus instead on the products of that behind-the-scenes work.

1234**5**6789101112 13

BASIC TYPES OF QUANTITATIVE COMPARISONS

One of the most fundamental skills in writing about numbers is describing the direction and magnitude of differences among two or more values. You may need to quantify the size of a difference—whether an election was close or a landslide, for example. You may want to assess the pace of change over time—whether population growth was slow or rapid in recent years, for instance. Or you may want to show whether a value exceeds or falls short of some important cutoff, such as whether a family's income is below the poverty level, and if so by how much.

There are several commonly used ways to compare numeric values: rank, difference, ratio, percentage difference, and z-scores.[1] Another measure—attributable risk—is less familiar but is a useful addition to your quantitative toolkit. With the exception of z-scores, which require some basic statistical knowledge, these calculations involve no more than elementary arithmetic skills—subtraction, multiplication, or division. For most authors, the difficult part of quantitative comparison is deciding which of those options is best suited to the question at hand, and then explaining the results and their interpretation clearly and correctly. In the sections below, I describe how to choose among, calculate, and describe the results of these calculations.

COORDINATING CALCULATIONS WITH PROSE

As with the tools described in the last few chapters, an important aspect of working with quantitative comparisons is coordinating them with the associated narrative. Think about how you prefer to word your comparisons, then perform the corresponding calculations accordingly. Doing so will spare your readers the confusion of interpreting ratios that you have accidentally described "upside down" or subtraction you have inadvertently explained "backward" (see examples below).

CHOICE OF A REFERENCE VALUE

The first step in comparing numbers is deciding which values to compare. Often you will contrast a figure from your own data against numbers or distributions from other sources. For instance, you might compare average children's height for one country against international growth standards, or today's scorching temperature against historic records for the same date and location. In addition, you might contrast several values from within your own data, such as average heights of children from several different countries, or the daily high temperatures over the past week.

Use of Standards and Conventions

As discussed in chapter 4, standards and conventions exist in many fields. A few examples:

- Since early 2003, the period 1982–1984 has been used as the base or reference period for the Consumer Price Index when calculating inflation (US Bureau of Labor Statistics 2007).
- National norms for exam scores and physical growth patterns are standard distributions used to evaluate individual test scores or height measurements.
- The Federal Poverty Thresholds (DeNavas-Walt, Proctor, and Smith 2013) are reference values that are commonly used to assess family income data.

If conventions or standards are commonly applied in your field and topic, use them as the reference values in your calculations.

Comparisons within Your Own Data

If external standards don't exist or you want to provide additional contrasts, pick salient comparison values within your own data. Some guidelines about deciding on a reference value or group.

- The value for all groups combined is a good comparison value for bivariate comparisons as long as none of the subgroups comprises too large a share of that whole. None of the 50 states is so large that its value dominates the total population of the United States, so the United States is a fine basis of comparison for each of the individual states. However, comparing values for males against those of both sexes combined is a poor choice because males make up half the population, strongly influencing the value for the whole population. Instead, compare males to another subgroup—in this case females.
- Social norms and considerations of statistical power often suggest comparison against the modal category. In language comparisons within the United States, English speakers comprise a sensible reference group.
- Pick a reference group to suit your audience and research question. For a report to your state's Department of Human Services, national values or those for adjacent states are logical points of reference. For a presentation to a construction workers' union, compare their benefits, salary, and occupational safety figures against those for other industries and occupations.
- If there is no standard benchmark date for a temporal comparison, changes are often calculated relative to the earliest value under consideration, sometimes against the most recent value.

Choice of a comparison value may depend on data availability.

- The US Census is conducted at 10-year intervals and surveys are conducted periodically, so you may have to compare against the closest census or survey date even if it doesn't exactly match the date of interest.
- Information on small groups or unusual cases may have to be approximated using data on related groups or similar cases.

The initial choice of a reference group or value may be arbitrary: it might not matter which group or place or date you choose as the basis of comparison. However, once you have selected your reference, be consistent, using it for all calculations addressing a particular question. If you have decided on the Midwest as the reference region, calculate and report values for each of the other regions compared to the Midwest, not the South versus the Midwest and the Northeast versus the West.

If you compare against a standard threshold or other value from outside your data, report its value in the text, table, or chart. For comparisons against standard distributions involving many numbers (e.g., national norms for test scores or physical growth) provide descriptive statistics or a summary chart, then refer to a published source for more detailed information.

Wording for Comparisons

Name the comparison or reference group or value in your description of all numeric contrasts so that the comparison can be interpreted. "The sex ratio was 75" doesn't convey whether that is 75 males per 100 females or 75 females per 100 males. The two meanings are not interchangeable.

Before you choose a reference value within your own data, anticipate how you will word the description. If you naturally want to compare all the other regions to the Midwest, make it the reference, then calculate and describe accordingly: "The Northeast is [measure of difference] larger (or smaller) *than the Midwest.*" Without the phrase "than the Midwest," it isn't clear whether the comparison is relative to the past (in other words, the region grew), to other concurrent estimates of the Northeast's population, or to some other unspecified region.

TYPES OF QUANTITATIVE COMPARISONS

There are several types of numeric contrasts, each of which provides a different perspective on the direction and magnitude of differences between numbers. In addition to reporting the values themselves, use rank, difference, ratio, percentage difference, z-score, or attributable risk to help interpret the meaning of those values in the context of your research question. In this chapter, I use those terms for convenience and to relate them to mathematical and statistical concepts used elsewhere. For most audiences, however, avoid technical jargon in your writing, substituting phrases such as those shown in the illustrative sentences throughout this chapter.

Value or Level

The value is the amount or level of the measure for one case or time point: e.g., the infant mortality rate (IMR) in the United States in 2014; the current cost of a gallon of gasoline in London. Always report the value, its context, and its units. However, reporting the value alone leaves its purpose and meaning unclear. Is this year's IMR higher or lower than last year's? Than other similar countries? By how much? Is that a lot? To answer such questions, include one or more of the following numeric comparisons.

Rank

Rank is the position of the value for one case compared to other values observed in the same time or place, to a well-established standard, or against a historic high or low value. How does this year's IMR in the United States compare to that in other countries in the same year? To its IMR for last year? To the lowest IMR ever observed anywhere? "Seventh lowest," "highest ever," and "middle of the pack" are examples of rankings. Two identical values share the same rank just as two identical race times or two equal vote tallies constitute a tie. For instance, the values of x for cases 4 and 5 in table 5.1 are both 50 (column 1), so they share the rank of 1 (column 2).

PERCENTILE

When many cases are being compared, use percentiles to convey rank. An assessment of a boy's height includes his percentile score to show how his height compares to that of all boys the same age. Percentiles are calculated by ranking all of the values and categorizing them into 100 groups each containing an equal share (1/100th) of the distribution (Utts 1999). Values that are lower than 99% of all other values are in the zeroth (or bottom) percentile, while those that exceed 99% of all other values are in the 99th (or top) percentile. The 50th percentile, or middle value, is the median; half of all values are lower and half are higher than the median. Because percentiles encompass all the values

Table 5.1. Formulas and case examples for different types of quantitative comparisons

Formula	(1) Reference value	(2) Rank based on x	(3) Number of interest	(4) Rank based on y	(5) Difference	(6) Ratio	(7) Percentage difference or percentage change
	x		y		$y - x$	y/x	$[(y - x)/x] \times 100$
Case 1	1	3	2	5	1	2.00	100
Case 2	1	3	26	3	25	26.00	2,500
Case 3	25	2	50	2	25	2.00	100
Case 4	50	1	25	4	-25	0.50	-50
Case 5	50	1	51	1	1	1.02	2

Note: Ranks based on x and y are from highest (1) to lowest.

in a distribution, they are bounded between 0 and 99: it is impossible to be below the lowest value or above the highest value.

To describe rank in somewhat less detail, report deciles (ranges of ten percentiles), quintiles (one-fifth of the distribution, encompassing 20 percentiles), quartiles (ranges of 25 percentiles), or terciles (the bottom, middle, and top third of the distribution).

WORDING FOR RANK

To report rank, select words that describe both relative position and the concept being compared: "fastest," "least expensive," and "second most dense" make it clear that the descriptions pertain to velocity, price, and density, respectively. If you use the words "rank" or "percentile," also mention what is being compared: "Kate's SAT score placed her in the second highest quartile nationwide," or "The United States ranked seventh best in terms of infant mortality."

Rank and percentile do not involve units of their own but are based on the units of the values being compared. If you have already reported the values and their units elsewhere, omit them from your description of rank. If not, rank and value can often be incorporated into the same sentence.

Poor: "The South's rank for population is 1."
  Although this statement may be correct, it can be restated in a
  more straightforward and informative way.
Better: "With a population of 114.6 million persons, the South was
  the largest region of the United States in 2010 (column 8 of
  table 5.2)."
  This version conveys the value, rank, and context in easily
  understood terms.

ADVANTAGES AND DISADVANTAGES OF RANK

Rank is useful when all that matters is the order of values, not the distance between them. In elections, the critical issue is who came in first. Rank in class or quartiles of standard test scores are often used as admission criteria by colleges.

Although rank and percentile both provide information about relative position (higher or lower, faster or slower), they do not indicate

Table 5.2. Application of rank, difference, ratio, and percentage change to US population data

Region	(1) Population (millions), 2000	(2) Region rank, 2000	(3) Ratio, 2000 (relative to Midwest)	(4) % Difference, 2000 (rel. to Midwest)	(5) Population (millions), 2010	(6) Population change (millions), 2000 to 2010	(7) % Change, 2000 to 2010	(8) Region rank, 2010
United States	281.4	NA	NA	NA	308.7	27.3	9.7	NA
Northeast	53.6	4	0.83	−16.8	55.3	1.7	3.2	4
Midwest	64.4	2	1.00	0.0	66.9	2.5	3.9	3
South	100.2	1	1.56	55.6	114.6	14.4	14.4	1
West	63.2	3	0.98	−1.9	71.9	8.7	13.8	2

Source: US Census Bureau 2010.

by how much one value differed from others. In the 2000 US presidential election, rank in electoral votes identified the winner but was not the only salient information. Bush's small margin of victory over Gore caused much debate and recounting of the popular vote—demands that would not have surfaced had the difference been larger. And some might argue that students at the bottom of the highest quartile and at the top of the second quartile of the SAT score distribution are so similar that they should receive similar college admission offers.

To quantify size of difference between two or more values, use difference, ratio, percentage difference, or *z*-score. Their formulas (except *z*-scores) are presented in table 5.1, along with some numeric examples. Throughout table 5.1, *x* represents the reference or comparison value, and *y* represents another number of interest.

Difference or Change

Difference[2] subtracts the reference value (*x*; column 1 of table 5.1) from the number of interest (*y*; column 3), or *y* minus *x*. For case 1 in table 5.1, the difference is 1 unit ($y - x = 2 - 1 = 1$; column 5). The difference for case 5 is also 1 unit ($51 - 50 = 1$), although both *x* and *y* are much higher. *Change* subtracts an earlier value from a more recent value, such as the US population in 2010 minus its population in 2000. Difference and change can be used for either interval or ratio variables.

WORDING FOR DIFFERENCE

A difference or change is computed by subtracting one value from another, so describe it in terms of a *difference* or *margin*. Mention the units, which are the same as those for the values being compared.

> *Poor:* "The change was 27.3 million (table 5.2)."
> *This sentence reports the magnitude but not the direction of the change and does not explain what concept or cases are being compared.*
> *Better:* "In 2010, the Census Bureau estimated a total US population of 308.7 million persons—an increase of 27.3 million over the population in 2000 (column 6 of table 5.2)."

This version specifies the two cases being compared (their years), mentions the pertinent concept (population) and units (millions of people), and reports the direction and size of the change.

ADVANTAGES AND DISADVANTAGES OF DIFFERENCE

The difference or change is useful when the difference itself is of interest: How much more will something cost, and is that amount within your budget? How many more people live in South Florida now than 10 years ago, and will that additional population overtax water supplies?

However, the difference does not address all questions well. Is an 8-ounce weight loss big? For a premature infant weighing only 3.5 pounds (56 ounces), an 8-ounce weight loss could be life-threatening. For a sumo wrestler weighing 350 pounds (5,600 ounces), an 8-ounce weight loss would hardly be noticeable.

Ratio

A ratio divides the number of interest (*y*) by the reference value (*x*). If the quantity in numerator (*y*) is larger than that in the denominator (*x*), the ratio is greater than 1.0, as in cases 1, 2, 3, and 5 in column 6 table 5.1. If the numerator is smaller than the denominator, the ratio is below 1.0, as in case 4 (25/50 = 0.50).

By dividing one value by the other, a ratio adjusts for the fact that a 1-unit difference has very different interpretations if both values are very small than if both values are very large. In both cases 2 and 3, the difference between *x* and *y* is 25 units. However, the ratio is much larger in case 2 (ratio = 26.0) because the reference value is very small (*x* = 1). In case 3, the reference value is much higher (*x* = 25), yielding a much smaller ratio (2.0). Ratios can be used for ratio-level variables but not interval-level variables: it makes no sense to say that it is 1.25 times as hot today as yesterday, for example.

WORDING FOR RATIOS

Describe ratios in terms of *multiples*: the value in the numerator is some multiple of the value in the denominator (without using that jargon; see table 5.3). Do not just report the ratio in the text without accompanying explanation:

Poor: "In 2000, the Southern numerator was 1.56 times the Midwestern denominator (column 3 of table 5.2)."
The terms "numerator" and "denominator" indicate that the comparison involves division, but the sentence doesn't express what aspect of the regions is being compared.

Poor (version 2): "In 2000, the ratio between the South and the Midwest was 1.56 (column 3, table 5.2)."
This version doesn't convey what is being measured or which region has the higher value.

Better: "In 2000, the South was 1.56 times as populous as the Midwest (column 3, table 5.2)."
In this version, it is immediately evident what is being compared, which region is bigger, and by how much.

If the ratio is close to 1.0, use wording to convey that the two values are very similar. For ratios below 1.0, explain that the value is smaller than the reference value. For ratios above 1.0, convey that the value is larger than the reference value. Table 5.3 gives examples of ways to explain ratios (including relative risks and odds ratios) without using phrases such as "ratio," "numerator," or "denominator."

COMMON ERRORS WHEN DESCRIBING RATIOS
Some cautions: word your explanation to conform to the kind of calculation you performed. I have seen people subtract to find a 2-unit difference between scores of 73 and 71, and then state that the score for the second group was twice as high as for the first. Likewise, if you divide to find a ratio of 1.03, do not explain it as a "1.03 unit difference" between the quantities being compared.

Explain ratios in terms of multiples *of the reference value*, not multiples *of the original units*. For example, although the populations of the South and Midwest were originally measured in millions of persons (column 1 of table 5.2), the ratio of 1.56 does not mean there were 1.56 million times as many people in the South as in the Midwest. During the division calculation the millions "cancel," as they say in fourth-grade math class, so there were 1.56 times as many people in the South as in the Midwest in 2000.

Avoid calculating the ratio with one group as the denominator

Table 5.3. Phrases for describing ratios and percentage difference

Type of ratio	Ratio example	Rule of thumb	Writing suggestion[a]
<1.0 (e.g., 0.x) *Percentage difference* = ratio × 100	0.80	[Group] is only x% as ___[b] as the reference value.	"Males were only 80% as likely as females to graduate from the program."
Close to 1.0	1.02	Use phrasing to express similarity between the two groups.	"Average test scores were similar for males and females (ratio = 1.02 for males compared to females)."
>1.0 (e.g., 1.y) *Percentage difference* = (ratio − 1) × 100	1.20	[Group] is 1.y times as ___ as the reference value.	"On average, males were 1.20 times as tall as females."
		or [Group] is y% ___er than the reference value.	*or* "Males were on average 20% taller than females."
	2.34	[Group] is (2.34 − 1) × 100, or 134% ___er than the reference value.	"Males' incomes were 134% higher than those of females."
Close to a multiple of 1.0 (e.g., z.00)	2.96	[Group] is (about) z times as ___.	"Males were nearly three times as likely to commit a crime as their female peers."

[a] Females are the reference group (denominator) for all ratios in table 5.3.
[b] Fill in each blank with an adjective, verb, or phrase to convey the aspect being compared, e.g., "tall," "likely to graduate."

and then explaining it "upside down": for example, do not report the relative size of the Southern and Midwestern populations in a table as ratio = 1.56 and then describe the comparison as Midwest versus South ("The population of the Midwest was 0.64 times that of the South").[3] Decide in advance which way you want to phrase the comparison, then compute accordingly.

Percentage Difference and Percentage Change

Percentage difference is a measure that expresses the difference between two values as a ratio compared to a specified reference value. As shown in column 7 of table 5.1, a 1-unit difference (from subtraction) yields a much larger percentage difference with an initial level of 1 (case 1 in table 5.1) than for an initial level of 50 (case 5). To compute percentage difference, divide the difference by the reference value, then multiply the result by 100 to put it in percentage terms: $[(y - x)/x] \times 100$, with the results in units of *percentage points*. If you do not multiply by 100, you have a *proportionate* difference.

Percentage difference is typically calculated by subtracting the smaller from the larger value, hence such comparisons often yield positive percentage differences. Negative percentage differences usually occur only when several values are being compared against the same reference value, with some falling below and some above that value. For example, in 2000 the Northeast was 16.8% smaller than the Midwest ($[(53.6$ million $- 64.4$ million)/64.4 million] $\times 100 = -16.8\%$), whereas the South was 55.6% larger than the Midwest ($[(100.2$ million $- 64.4$ million)/64.4 million] $\times 100 = 55.6\%$; column 4, table 5.2).

A percentage *change* compares values for two different points in time. Conventionally, a percentage change subtracts the earlier (V_1) from the later value (V_2), then divides that difference by the initial value and multiplies by 100: $[(V_2 - V_1)/V_1] \times 100$.

- If the quantity increased over time, the percentage change will be positive. For the West region: $[(V_{2010} - V_{2000})/V_{2000}] \times 100 = [(71.9$ million $- 63.2$ million)/63.2 million] $\times 100 = [8.7$ million/63.2 million] $\times 100 = 13.8\%$, reflecting a 13.8% increase in population between 2000 and 2010 (column 7 of table 5.2).

- If the quantity decreased over time, the percentage change will be negative. Between 2000 and 2010, the state of Michigan lost 0.6% of its population, decreasing from 9,938,444 persons to 9,883,640 in 2010 (US Census Bureau 2010): $[(9,883,640 - 9,938,444)/9,938,444] \times 100 = [(-54,804/9,938,444)] \times 100 = -0.6\%$.

- If the time interval is very wide (e.g., several decades or centuries), sometimes the average of the values for the two times is used as the denominator: $[(V_2 - V_1)/(V_1 + V_2)/2] \times 100$. When you report a percentage change, indicate which date or dates were used as the base for the calculation.

A percentage difference is a variant of a ratio: if you know either the ratio or percentage difference between two values, you can calculate the other measure, as explained in table 5.3:

- For ratios that are greater than 1.0, percentage difference = (ratio − 1) × 100. Conversely, ratio = (percentage difference/100) + 1. An article in the *American Journal of Public Health* reported that ready-to-eat cookies being sold in some popular fast-food and family restaurants have 700% more calories than the standard USDA portion size (Young and Nestle 2002)—a ratio of eight times the calories of a "standard" cookie.
- For ratios less than 1.0, percentage difference = ratio × 100. If there are 0.83 Northeasterners per Midwesterner, then the population of the Northeast is 83% as large as that of the Midwest (column 3, table 5.2).

WORDING FOR PERCENTAGE DIFFERENCE OR CHANGE

To describe percentage difference or percentage change, identify the cases being compared and the direction of the difference or change. A percentage difference is expressed as a percentage of the reference value, replacing the units in which the original values were measured. A percentage change across time is expressed as a percentage of the earlier value.

> *Poor:* "The Western population percentage change was 13.8."
> *This sentence is awkwardly worded and does not convey which dates are being compared.*
> *Better:* "In the first decade of the twenty-first century, the population of the West region grew from 63.2 million to 71.9 million persons—an increase of 13.8% (column 7, table 5.2)."

> *This version reports both of the values (population in each time period) and the percentage change, including direction, magnitude, concepts, and units.*

To report a negative value of percentage change or percentage difference, select words to convey direction:

Poor: "In 2000, the populations of the Northeast and Midwest were 53.6 million and 64.4 million persons, respectively, so the percentage difference between the West and the Midwest is negative (−16.8%; column 4, table 5.2)."
Although this sentence reports the correct calculation, wording with negative percentage differences is unwieldy.

Better: "In 2000, the Northeast had 16.8% fewer inhabitants than the Midwest (53.6 million persons and 64.4 million persons, respectively; table 5.2)."
The phrase "fewer inhabitants than the Midwest" clearly explains the direction of the difference in population between regions.

COMMON ERRORS FOR WORDING OF PERCENTAGE DIFFERENCE
Do not confuse the phrases "*y* is 60% *as high as* *x*" and "*y* is 60% *higher than* *x*." The first phrase suggests that *y* is lower than *x* (i.e., that the ratio *y/x* = 0.60), the second that *y* is higher than *x* (i.e., *y/x* = 1.60). After you calculate a ratio or percentage difference, explain both the direction and the size of the difference, then check your description against the original numbers to make sure you have correctly communicated which is bigger—the value in the numerator or that in the denominator. See also table 5.3 for example sentences.

Watch your math and reporting of units to avoid mixing or mislabeling percentages and proportions. A proportion of 0.01 equals 1%, not 0.01%.

Other Related Types of Quantitative Comparison
ANNUAL GROWTH RATES
Many annual rates such as interest rates or population growth rates are measures of change over time that are reported in percentage form. However, an annual growth rate cannot be calculated simply by

dividing a percentage change over an n-year period by n. For example, the annual growth rate in the West between 2000 and 2010 was *not* 13.8%/10 years. Box 5.1 illustrates the effect of annual compounding in a savings account over a 10-year period, in which interest is added to the principal each year, so the interest rate is applied to a successively larger principal each year. The same logic applies to population growth: each year there are more people to whom the annual growth rate applies, so even if that growth rate is constant, the number of additional people rises each year.[4] Between 2000 and 2010, the West region grew at an average annual rate of 1.29%.[5]

PERCENTAGE VERSUS PERCENTILE VERSUS PERCENTAGE CHANGE

A common source of confusion involves "percentage difference," "difference in percentage points," and "difference in percentiles." Just because they all have "percent-" in their names does not mean they are equivalent measures. If x and y are expressed as percentages, their units of measurement are percentage points; hence the difference between their values is reported as a *difference in percentage points*. A rise in the unemployment rate from 4.2% to 5.3% corresponds to an increase of 1.1 percentage points, not a 1.1% difference. The *percentage difference* in those unemployment rates is (5.3% − 4.2%)/4.2% × 100, which is a 26% increase relative to the initial value.

Percentages and percentiles calculate the share of a whole and the rank within a distribution, respectively. By definition, neither can be less than zero or greater than 100: no case can have less than none of the whole, or more than all of it. In contrast, *percentage change* and *percentage difference* measure relative size against some reference value and are not constrained to fall between 0 and 100. If a value is more than twice the size of the reference value, the percentage difference will be greater than 100%, as in case 2 in table 5.1. Similarly, if the quantity more than doubles, the corresponding percentage change will exceed 100%. If a quantity shrinks over time, as did the population of Michigan between 2000 and 2010, the corresponding percentage change will be less than 0% (negative).

To illustrate how percentage, percentile, and percentage change interrelate, box 5.2 and table 5.4 apply those measures to SAT scores for a fictitious student. The description also illustrates how to integrate several different types of quantitative comparison into a coherent discussion to make different points about the numbers.

BOX 5.2. RELATIONS AMONG PERCENTAGE, PERCENTILE, AND PERCENTAGE CHANGE

The following description is annotated to indicate whether the numbers are value (denoted "V"), percentage correct (P), rank (R), difference (D), or percentage change (C). Those annotations and the material in brackets are intended to illustrate how the different statistics relate to one another, and would be omitted from the description for most audiences.

"The first time he took the SATs, Casey Smith correctly answered 27 out of 43 (V) questions (63%) (P; table 5.4). Compared to all students who took the test nationwide, he placed in the 58th percentile (R). The next year, he improved his score by 9 percentage points (D) [from 63% of questions correct (P) to 72% correct (P)], placing him in the 70th percentile (R). That change was equivalent to a 14.3% improvement in his score (C) [a 9 percentage-point improvement (D), compared to his initial score of 63% correct (V)]. His rank improved by 12 percentiles (D) [relative to his first-year rank of 58th percentile (R)]."

Table 5.4. Examples of raw scores, percentage, percentile, and percentage change

Comparison of standardized test scores, Casey Smith, 2013 and 2014

	2013	2014
Number of questions correct (V)[a]	27	31
Total number of questions (V)	43	43
Percentage of questions correct (P)	63%	72%
Difference in % correct (vs. 2009) (D)	NA	9%
Percentile (compared to national norms) (R)	58	70
Percentage change in % correct (vs. 2009) (C)	NA	14%

[a] Letters in parentheses coordinate with abbreviations in box 5.2.

Standardized Score or z-Score

Standardized scores, or z-scores, are a way of quantifying how a particular value compares to the average, taking into account the spread in the sample or a reference population (Agresti and Finlay 1997). A z-score for the ith case is computed by subtracting the mean (μ) from the value for that case (x_i), then dividing that difference by the standard deviation (σ), or $z = (x_i - \mu)/\sigma$. A positive z-score corresponds to a value above the mean, a negative z-score to a lower-than-average value. For instance, on a test with a mean of 42 points and a standard deviation of 3 points, a raw score of 45 points corresponds to a z-score of 1.0, indicating that the individual scored one standard deviation above the mean. A z-score is best used when the distribution is approximately normal (bell-shaped). Standardized scores are often used to provide a common metric that allows variables with very different ranges to be compared with one another.

In addition to correcting for level by subtracting the mean value, z-scores adjust for the fact that a given difference is interpreted differently depending on the extent of variation in the data. Among six-month-old male infants in the United States in 1999, mean height was 66.99 centimeters (cm) with a standard deviation (σ) of 2.49 cm (Centers for Disease Control and Prevention 2002). Hence a six-month-old boy who was 2.54 cm shorter than average would have a z-score of −1.02, indicating height roughly one standard deviation below the mean. Among six-year-old boys, however, there is wider variation around mean height (115.39 cm; σ = 5.05), so a boy 2.54 cm shorter than average (the same difference as for the infant) would be only about half a standard deviation below the norm for his age ($z = -0.50$).

Sometimes the mean and standard deviation used to calculate z-scores are from within your sample, other times from a standard population. For example, international growth standards for children's height-for-age and weight-for-height are used to provide a consistent basis for evaluating prevalence of underweight, overweight, and long-term growth stunting in different populations (World Health Organization 1995; Kuczmarski et al. 2000).

Report the mean and standard deviation of the distribution you used to derive the z-scores, and mention whether those figures were derived from within your own data or from an external standard. If an external standard was used, name it, explain why it was chosen, and provide a citation. If you report z-scores for only a few selected cases, also report the unstandardized values for those cases in the text or a table; for more cases, create a graph comparing your sample to the reference distribution.

The units of a z-score are *multiples of standard deviations*, not the original units in which the variable was measured: "With a raw score of 72, Mia scored one standard deviation below the national average."

Attributable Risk

An important issue for many research questions is the substantive significance or "real-world" meaning of the pattern under study (see chapter 3). Imagine that you oversee the Superfund cleanup effort and must decide which waste sites to clean up first, given a limited budget. Most people assume that priority automatically should go to removing materials that have a high relative risk[6]—those that drastically increase health risks. However, another important but often ignored determinant of the potential impact of a risk factor (predictor variable) is its prevalence—how common it is—in this case measured by the proportion of the population exposed to that material.

To measure the net impact of both prevalence and relative risk, epidemiologists use a calculation called attributable risk, also variously referred to as "attributable fraction" and "population attributable risk." Attributable risk can be thought of as the maximum percentage reduction in the incidence of the outcome if no one were exposed to the suspected risk factor (Lilienfeld and Stolley 1994), and is used to compare the burden of different diseases and preventable risk factors (e.g., World Health Organization 2002). For example, what fraction of cancer cases could be prevented if all exposure to a certain toxic substance were eliminated? What share of low birth weight cases could be averted if no women smoked while they were pregnant?

Although it is little known outside of epidemiology, attributable risk also can shed light on other kinds of research questions. Suppose more students fail a proficiency test under the current math curriculum than under a newer curriculum. Attributable risk calculations can be used to estimate the percentage of failing scores that could be eliminated if the better curriculum completely replaced the current one.

Attributable risk is calculated $AR = [p \times (RR - 1)]/[(p \times [RR - 1]) + 1] \times 100$, where RR is the relative risk of the outcome for those exposed to the risk factor versus those not exposed, and p is the proportion of the population exposed to the risk factor. Specify the lowest risk category as the reference group of the predictor variable for the relative risk calculation. In the math curriculum example, RR measures the relative risk of failing the exam for students using the current compared to the new math curriculum, and p is the proportion of students using the current math curriculum.

Table 5.5 reveals that both relative risk and prevalence of the risk factor have substantial influences on attributable risk. For instance, an attributable risk of 47% can be obtained by either the combination (RR = 10.0 and p = 0.10) or (RR = 2.0 and p = 0.90), two very different scenarios. Risk factors that are both very common and have large relative risks have the largest attributable risk, while those that are both uncommon and have modest effects have the smallest impact (e.g., RR = 2.0, p = 0.10 yield an AR of 9%).

The logic behind attributable risk assumes a causal relationship: that eliminating the risk factor will reduce the probability of the outcome. Consequently, use relative risk estimates based on data from a randomized experiment, or—if you have observational data—from a multivariate model that controls for potential confounding factors. With those caveats in mind, attributable risk is a compelling statistic

Table 5.5. Relationship between relative risk, prevalence of a risk factor, and attributable risk

Attributable risk (%) for selected values of relative risk (RR) and proportion of the population with the risk factor (p)

p = Proportion of population exposed to risk factor	Attributable risk (%)		
	RR = 2.0	RR = 4.0	RR = 10.0
0.10	9	23	47
0.25	20	43	69
0.50	33	60	82
0.75	43	69	87
0.90	47	73	89

for journalists and policy makers because it quantifies maximum potential benefits of a proposed solution.

Provide data on the prevalence of each risk factor and its associated relative risk either in tables or in the text, and specify the reference group for the relative risk calculations. When comparing attributable risk for several different risk factors, create a table to present relative risk, prevalence, and attributable risk (columns) for each risk factor (rows). For single attributable risk calculations in a results or discussion section, report that information in the text, table, or a footnote.

Poor: "The attributable risk was 9% (RR = 1.4; prevalence = 0.25)."
 This version leaves the units ("percentage of what?") and interpretation of attributable risk unexplained, and fails to mention the pertinent risk factor ("prevalence of what?") or outcome ("relative risk of what?").

Better: "Approximately 25% of pregnant women in the United States smoke. Combined with the estimated 1.4-fold increase in odds of low birth weight associated with maternal smoking, this figure suggests that if all pregnant women could be persuaded not to smoke, roughly 9% of low birth weight cases could be eliminated."

Attributable risk is explained in terms of familiar concepts, avoiding introduction of a term that would be used only once. The statistics used in the calculation are cited in the text and units for each number are clearly specified.

For scientific readers who are unfamiliar with attributable risk, include a footnote such as:

"Based on attributable risk calculations: AR = $[p \times (RR - 1)]/[(p \times [RR - 1]) + 1] \times 100$, where relative risk (RR) = 1.4 and prevalence of smoking (p) = 0.25 (see Lilienfeld and Stolley 1994)."

This footnote names the measure (attributable risk) and gives a reference for additional information. It also relates components of that calculation to specific numeric values and concepts for this research question.

CHOOSING TYPE(S) OF QUANTITATIVE COMPARISONS FOR YOUR WRITING

The types of variables you are working with constrain which types of quantitative comparisons are appropriate (Chambliss and Schutt 2012).

- Ratio variables: rank, difference, ratios, and calculations involving more than one of those operations all make sense because 0 is the lowest possible value.
- Interval variables: rank and difference work, but ratios do not because such variables can take on both positive and negative values.
- Ordinal variables: only rank can be calculated.
- Nominal variables: the only pertinent comparison is whether different cases have the same or different values, but differences cannot be quantified or ranked.

For variables where several of these contrasts are possible, different types of quantitative comparisons provide distinct perspectives about the same pair of numbers. Always report the value to set the context and provide data for other calculations, then present one or two types of comparisons to give a more complete sense of the relationship. To help readers interpret both value and difference, mention the highest and lowest possible values and the observed range in the data. A one-point increase on a five-point Likert scale is substantial—equal to one-fourth of the theoretically possible variation. A one-point increase in the Dow Jones Industrial Average is minuscule—equivalent to less than a 0.01% change compared to its level of nearly 16,000 points in early 2014.

The value is also important for putting a ratio in context: suppose a study reports that a pollutant is three times as concentrated in one river as in another. A very low concentration in both locations (e.g., 1 part per million [ppm] versus 3 ppm) has very different environmental implications than a high concentration in both (e.g., 1% versus 3%). Likewise, reporting the percentage difference or percentage change without the value can be misleading. A 100% increase (doubling) in the number of scooter-related injuries over a three-month period might be considered alarming if the injury rate in the baseline period was already high, but not if there were initially few injuries because scooters were not in widespread use.

The choice of which calculations to include depends on your topic and discipline. Read materials from your field to learn which types of quantitative comparisons are customary in your field. Use those calculations, then consider whether other types of comparisons would add valuable perspective.

- Report results of a race, election, or marketing study in terms of rank and difference.
- Describe time trends in terms of difference and percentage change, substituting ratios to express large changes such as doubling, tripling, or halving.
- Describe variations in risk or probability in terms of ratios. Because the ratio and percentage change are simply mathematical transformations of one another (table 5.3), present only one of those statistics to avoid repetition.
- Likewise, report only one measure of rank (e.g., position, percentile, decile, or quartile) along with information on the number of cases involved in the ranking.

- Convey potential reductions in risk associated with eliminating a risk factor using attributable risk.

As you write about results of quantitative comparison, specify the type of measure used as the input values (data) for those calculations so your readers know the basis of your conclusions. For instance, rankings of car thefts are quite different depending on which measure was used to create the rankings: In 2011, the most frequently stolen car in terms of sheer number of vehicles (and thus as a percentage of all car thefts) was the Honda Civic (NICB 2011), probably because there are so many Civics on the road. However, the car with the highest theft *rate* was the Chevrolet Corvette, with 10% of that model ever manufactured stolen (NICB 2012). These statistics suggest that although the Civic was the car model most *frequently* or most *often* stolen, the Corvette was the model most *likely* (highest risk) to be stolen.

CHECKLIST FOR TYPES OF QUANTITATIVE COMPARISONS

- Always report the units and values of the main numbers being compared, either in the text itself or in a table or chart.
- Select one or two additional types of quantitative comparisons.
- Specify which time, place, or group is the reference value.
- Interpret your calculations to convey whether the values are typical or unusual, high or low. For trends, explain whether the values are stable or changing, and in what direction.
- Describe results of each quantitative comparison to match its calculation:
    "Difference" or "margin" of the original units for subtraction
    Multiples of the reference value for ratios
    Multiples of standard deviations for z-scores
    Percentage reduction in risk compared to the reference group for attributable risk; name the reference group, outcome, and risk factor in your explanation
- Indicate which measure was used in your calculations, e.g., whether you compared numbers of cases, rates, etc. when computing a ranking, difference, ratio, or percentage difference.
- Explain standard definitions, constants, or reference data, and provide citations.

CREATING EFFECTIVE TABLES

Good tables complement your text, presenting numbers in a concise, well-organized way to support your description. Make it easy for your audience to find and understand numbers within your tables. Design table layout and labeling that are straightforward and unobtrusive so the attention remains on the substantive points to be conveyed by your data rather than on the structure of the table. In this chapter, I explain the following:

- How to create tables so readers can identify the purpose of each table and interpret the data simply by reading the titles and labels
- How to make tables self-contained, including context, units, category names, source of the data, types of statistics, and definitions of abbreviations
- How to design a layout that contributes to the understanding of the patterns in the table and coordinates with your written description

The first section gives principles for planning effective tables. The second explains the "anatomy of a table"—the names and features of each table component. The third describes common types of tables, and the fourth gives guidelines on how to organize tables to suit your audience and objectives. The final sections offer advice about how to draft a table and create it on a computer.

PRINCIPLES FOR PLANNING EFFECTIVE TABLES
Creating Focused Tables
Many reports and papers include several tables, each of which addresses one aspect of the overall research question—one major topic or a set of closely related subtopics, or one type of statistical analysis, for instance. A consultant's report on different options for a new commuter rail system might present information on projected ridership under different fare pricing scenarios (one table); estimated costs of land acquisition for right-of-way, commuter parking, and other related needs (a second table); and annual financing costs under several different assumptions about interest rates and term of the loan (a third table).

Creating Self-Contained Tables
Often tables are used separately from the rest of the document, either by readers in a hurry to extract information or as data sources that become detached from their origins. Label each table so your audience can understand the information without reference to the text. Using the title, row and column headings, and notes, they should be able to discern the following:

- The purpose of the table
- The context of the data (the W's)
- The location of specific variables within the table
- Units of measurement or categories for every number in the table
- Data sources
- Definitions of pertinent terms and abbreviations

The units and sources of data can be specified in any of several places in the table depending on space considerations and whether the same information applies to all data in the table (see next section).

ANATOMY OF A TABLE
Title
Write a title for each table to convey the specific topics or questions addressed in that table. In documents that include several tables or

charts, create individualized titles to differentiate them from one another and to convey where each fits in the overall scheme of your analysis.

TOPIC

In the title, name each of the major components of the relationships illustrated in that table. To avoid overly long titles, use summary phrases or name broad conceptual categories such as "demographic characteristics," "physical properties," or "academic performance measures" rather than itemizing every variable in the table. (The individual items will be labeled in the rows or columns; see below.) The title to table 6.1 mentions both the outcome (number of households) and the comparison variables (household type, race, and Hispanic origin).

CONTEXT

Specify the context of the data by listing the W's in the table title: where and when the data were collected, and if pertinent, restrictions on who is included in the data (e.g., certain age groups). If the data are from a specific study (such as the National Survey of America's Families or the Human Genome Project) or institution (e.g., one college or hospital), include its name in the title or in a general note below the table. Minimize abbreviations in the title. If you must abbreviate, spell out the full wording in a note.

UNITS

State the units of measurement, level of aggregation, and system of measurement for every variable in the table. This seemingly lengthy list of items can usually be expressed in a few words such as "price ($) per dozen," or "distance in light-years." Whenever possible, generalize units for the table rather than repeating them for each row and column. If the same units apply to most numbers in the table, specify them in the title. If there isn't enough space in the title, or if the units vary, mention units in the column or row headings. The title to table 6.1 states that all statistics are reported as number of thousands of households. Tables of descriptive statistics often involve different units for different variables; in those instances, specify the

Table 6.1. Anatomy of a table

Households (thousands) by type, race, and Hispanic origin, United States, 1997

Characteristic	All households	Family households		Other families		Nonfamily households		
		Total	Married couple	Female householder	Male householder	Total	Female householder	Male householder
Race/ethnicity								
White	86,106	59,511	48,066	8,308	3,137	26,596	14,871	11,725
Non-Hispanic White	77,936	52,871	43,423	6,826	2,622	25,065	14,164	10,901
Black	12,474	8,408	3,921	3,926	562	4,066	2,190	1,876
All other[a]	3,948	2,961	2,330	418	212	986	455	532
Origin								
Non-Hispanic	93,938	63,919	49,513	11,040	3,366	30,018	16,762	13,258
Hispanic[b]	8,590	6,961	4,804	1,612	545	1,630	754	875
Total	102,528	70,880	54,317	12,652	3,911	31,648	17,516	14,133

Source: US Census Bureau 1998.

[a] "All other" races includes Asians, Pacific Islanders, Native Americans, and those of unspecified race.

[b] People of Hispanic origin may be of any race.

units for each variable in the associated column or row headings (see related sections below).

If only one type of statistic is reported in the table, mention it in the table title.

- For a univariate table, state whether it reports distribution or composition, mean values, or other descriptive statistics.
- For a bivariate table, indicate whether it reports correlations, differences in means, cross-tabulations, or other measure of association.

For tables that include several types of statistics, provide a summary moniker in the title, and identify the types of statistics in column, row, or panel headings.

USE OF SAMPLING WEIGHTS
If some or all of the statistics in a table are weighted, state so in the table title or a footnote and cite a reference for the source of the weights.

The following examples illustrate these principles for writing good titles:

Poor: "Descriptive statistics on the sample."
  This title is truly uninformative. What kinds of descriptive statistics? On what variables? For what sample?
Better: "Means and standard deviations for soil components"
  This title makes it clear what topics and statistics the table includes, but omits the context.
Best: "Means and standard deviations for soil components, 100 study sites, Smith County, 1990."
  In addition to mentioning the topic, this title mentions the date and place the data were collected, as well as sample size.

Row Labels
Name the concept for each row and column in its associated label so readers can interpret the numbers in the interior cells of the

table. The identity and meaning of the number in the most heavily shaded cell of table 6.1 is households of all types (known from the column header) that include black persons (row label = name of category in the row), with population measured in thousands of households (title).

If the units of measurement differ across rows or columns of a table, mention the units in the pertinent row or column label. A table of descriptive statistics for a study of infant health might include mean age (in days), weight (in grams), length (in centimeters), and gestational age (in weeks). With different units for each variable, the units cannot be summarized for the table as a whole. Do not assume that the units of measurement will be self-evident once the concepts are named: without labels, readers might erroneously presume that age was measured in months or years, or weight and length reported in British rather than metric units.

Minimize use of abbreviations or acronyms in headings. If space is tight, use single words or short phrases. Explain the concepts measured by each variable as you describe the table so the brief labels will become familiar. Do not use "alphabet soup" variable names from statistical packages—your audience will not know what they mean. Likewise, avoid using numeric codes as row labels for categories—your readers will not be using your database so they don't need to know those codes. Instead, label the categories with intuitive names, such as "Hispanic" or "Excellent health."

Consider the row labels in the following tables:

Poor:

TABLE X. DESCRIPTIVE STATISTICS ON SAMPLE

Variable	Mean	Standard deviation
YNDSR	##	##
Q201a	##	##

Short, cryptic acronyms such as "YNDSR" might be required by your statistics software but rarely are sufficient to convey the meaning of the variable. A variable name based on question number (such as "Q201a") can help you (the data analyst) remember which questionnaire item was the original source of the data, but obscures the concept measured by that variable. Feel free to use such shorthand in your data sets and in initial drafts of your tables, but replace them with meaningful phrases in the version your audience will see.

Better:

TABLE X. MEANS AND STANDARD DEVIATIONS ON SOCIOECONOMIC AND ATTITUDINAL VARIABLES

Variable	Mean	Standard deviation
Income-to-needs ratio	##	##
Extent of agreement with current welfare system	##	##

These labels clearly identify the concepts in each row. In a scientific article, define the income-to-needs ratio and attitudinal measures in the data and methods section or note to the table. For a lay audience, replace technical labels with everyday synonyms.

If readers need to see the long or complex wording of a question to understand the meaning of a variable or items in a scale or index, refer them to an appendix that contains the pertinent part of the original data collection instrument; see table 10.1 for an example.

INDENTING

When organizing rows in a table, place categories of a nominal or ordinal variable in consecutive rows under a single major row header, with the subgroups indented. Counts and percentages for subgroups that are indented equally can be added together to give the total. In table 6.1, for example, "White," "Black," and "All other" together comprise all households. Row labels that are indented farther indicate subgroups and should not be added with the larger groups to avoid double counting. "Non-Hispanic white" is indented in the row below "White," showing that the former is a subgroup of the latter. In the terminology of chapter 4, "white" and "non-Hispanic white" are not mutually exclusive, so they should not be treated as distinct groups when calculating totals or frequency distributions.[1] To indicate that Hispanics should not be added to the other racial groups within the table, the Hispanic origin contrast is given a separate left-justified row label with rows for Hispanics and non-Hispanics below. Finally, a footnote explains that Hispanics can be of any race, indicating that they should not be added to the racial categories.

PANELS

Use panels—blocks of consecutive rows within a table separated by horizontal lines ("rules") or an extra blank row—to organize material within tables. Arrange them one above another with column headings shared by all panels. Panels can introduce another dimension to a table, show different measures of the relationship in the table, or organize rows into conceptually related blocks.

Adding a dimension to a table

Examples of tables that use panels to introduce an additional dimension to a table:

- Separate panels for different years. For example, the relationship between race, ethnic origin, and household structure (table 6.1) might be shown at 10-year intervals from 1970 through 2010 with each year in a separate panel, labeled accordingly. The panels introduce a third variable to a two-way (bivariate) table, in this case adding year to a cross-tabulation of race/origin with household structure. The panels would share the column headings (household structure), but repeat the row headings (for race and ethnic origin) in each panel.
- Separate panels for other characteristics. For instance, the relationship between household structure and race and ethnic origin might be shown for each of several regions or income levels.

Different measures for the same relationship

Also use panels to organize a table that presents different measures of the same concept, such as number of cases or events, along with measures of distribution or rate of occurrence:

- Table 6.1 could be modified to include a second panel reporting the *percentage* of households in each category to supplement the first panel showing the *number* of households.

- A table might present *number of deaths* according to cause or other characteristics in one panel and *death rates* in another, as in many Centers for Disease Control and Prevention reports.

For these types of applications, repeat the row headings within each panel and specify the Ws and units separately in a header for each panel.

Organizing conceptually related blocks

When a table contains many related variables in the rows, use panels to organize them into blocks of similar items. Rather than lump all 10 questions on AIDS transmission into one section, table 6.2 is arranged into two panels—the top panel on knowledge of ways AIDS is likely to be transmitted, the bottom panel on ways it is unlikely to be transmitted—each labeled accordingly. Within each panel, results are shown separately for each specific question, followed by a summary statistic on that broad knowledge area.

If two small, simple tables have the same column headers and address similar topics, you can combine them into a single table with panels, one panel for the set of rows from each of the smaller tables. Although table 6.2 could have been constructed as two separate tables—one on likely modes of AIDS transmission and the other on unlikely modes—creating a single table facilitates comparison across topics, such as pointing out that the likely modes are all better understood than the unlikely modes.

For tables that you describe in your text, avoid using more than two or three panels per table, and try to fit each table onto one page or on facing pages. Refer to each panel in your written description to direct readers to specific concepts and numbers as you mention them. Appendix tables that organize data for reference use can include more panels and spill onto several pages. For multipage tables, repeat the table number and column headings on each page, and label the panels (topics, units, context) so that your readers can follow them without written guidance.

Column Headings

Each column heading identifies the variable or statistic (e.g., mean, standard deviation) in that column. The guidelines listed above for

Table 6.2. Use of panels to organize conceptually related sets of variables

Knowledge about AIDS transmission, by language spoken at home and ability to speak English, New Jersey, 1998

Mode of transmission	Language spoken at home/language used on questionnaire			Chi-square	p-value
	English (N = 408)	Spanish/ English ques. (N = 32)	Spanish/ Spanish ques. (N = 20)		
Likely modes of transmission[a]					
Sexual intercourse with an infected person	93.6	87.5	95.0	1.9	(.39)
Shared needles for IV drug use	92.4	90.6	65.0	17.6	(.000)
Pregnant mother to baby	89.5	75.0	80.0	7.2	(.03)
Blood transfusion from infected person	87.5	81.3	60.0	12.5	(.002)
Mean percentage of "likely" questions correct	91.7	83.6	75.0	8.3[b]	(.000)
Unlikely modes of transmission[a]					
Working near someone with the AIDS virus	81.6	75.0	35.0	25.4	(.000)
Using public toilets	66.4	53.1	30.0	12.7	(.002)
Eating in a restaurant where the cook has AIDS	61.3	50.0	35.0	3.7	(.04)
Being coughed or sneezed on	57.8	50.0	25.0	8.8	(.01)
Sharing plates, cups, or utensils	56.4	46.9	25.0	8.3	(.02)
Visiting an infected medical provider	35.0	34.4	25.0	0.8	(.65)
Mean percentage of "unlikely" questions correct	59.8	51.6	29.2	8.2[b]	(.000)
Mean percentage of all questions correct	72.1	64.4t	47.5	11.7[b]	(.000)

Source: Miller 2000a.

[a] Percentage of respondents answering AIDS transmission questions correctly.

[b] Test for difference based on ANOVA. Reported statistic is the *F*-statistic with 2 degrees of freedom.

labeling abbreviations, notes, and units and categories in rows also apply to columns. If most numbers in a large table are measured in the same unit, use a spanner across columns to generalize with a phrase such as "percentage unless otherwise specified," then name the units for variables measured differently (e.g., in years of age or price in dollars) in the pertinent column heading.

COLUMN SPANNERS

Column spanners (also known as "straddle rules") show that a set of columns is related, much as indenting shows how a set of rows is related. In table 6.1, households fall into two broad categories—family households and nonfamily households—each of which is demarcated with a column spanner. Beneath the spanners are the associated household subtypes: "Family households" comprise "Married couple" and "Other families," with "Other families" further subdivided into "Female householder" and "Male householder." Nonfamily households include those headed by a "Female householder" and those headed by a "Male householder." Each column spanner also encompasses a column for the total number of households of that type: the "Total" column under the "Family households" spanner is the sum of the "Married couple" and the two "Other families" columns.

Interior Cells

Report your numbers in the interior cells of the table, following the guidelines in table 4.2 for number of digits and decimal places. Many disciplines omit numeric estimates based on only a few cases, either because of the substantial uncertainty associated with those estimates or to protect confidentiality of human subjects (appendix 1 in Benson and Marano 1998; NCHS 2002). Conventions about minimum sample sizes vary by discipline, so follow the standards in your field. If there is an insufficient number of cases to report data for one or more cells in your table, type a symbol in place of the numeric estimate and include a footnote that specifies the minimum size criterion and a pertinent citation.

Notes to Tables

Put information that does not fit easily in the title, row, or column labels in notes to the table. Spell out abbreviations, give brief definitions, state use of sampling weights, and provide citations for data sources or other background information. To keep tables concise and tidy, limit notes to a simple sentence or two, referring to longer descriptions in the text or appendixes if more detail is needed. If a table requires more than one note, label them with different symbols or letters rather than numbers (which could be confused with exponents), then list the notes in that order at the bottom of the table following the conventions for your intended publisher. Labeling the notes with letters also allows the reader to distinguish table notes from text notes.

If you are using secondary data, provide a note to each table citing the name and date of the data set or a reference to a publication that describes it. If all tables in your article, report, or presentation use data from the same source, you might not need to cite it for every table. Some journals or publishers require the data source to be specified in every chart or table, however, so check the applicable guidelines.

COMMON TYPES OF TABLES

This section describes common variants of univariate, bivariate, and three-way tables. See also Nicol and Pexman (1999) for guidance on tables to present specific types of statistics and Miller (2013a) for multivariate tables.

Univariate Tables

Univariate tables show information on each variable alone rather than associations among variables. Common types of univariate tables include those that present the distribution of a variable or composition of a sample (table 6.3) or descriptive statistics on the key predictor and outcome variables for your study (table 6.4).

A univariate table can include more than one type of numeric information for each variable. Table 6.3 includes separate columns for the number and percentage of cases with each categorical attribute, labeled accordingly. Table 6.4 presents the mean, standard deviation, minimum, and maximum values for birth weight and several maternal characteristics measured as continuous variables.

A table can also be used to compare composition of a sample and target population (or "universe"), as in table 6.5, where characteristics

Table 6.3. Univariate table: Sample composition for categorical variables
Demographic characteristics of study sample, Faketown, 2013

Demographic characteristic	Number of cases	Percentage of sample
Gender		
Male	1,000	48.6
Female	1,058	51.4
Age group (years)		
18–39	777	37.8
40–64	852	41.4
65+	429	20.8
Education		
<High school	358	17.4
=High school	1,254	60.9
>High school	446	21.7
Race/ethnicity		
Non-Hispanic white	1,144	55.6
Non-Hispanic black	455	22.1
Hispanic	328	15.9
Asian	86	4.2
Other race	45	2.2
Overall sample	2,058	100.0

Table 6.4. Univariate table: Descriptive statistics for continuous variables
Descriptive statistics on infant health and maternal characteristics, 1988–1994 NHANES III

	Mean	Standard deviation	Minimum	Maximum
Birth weight (grams)	3,379.2	609.16	397	5,896
Age of mother at child's birth (years)	26.0	5.5	11	49
Mother's education (years)	12.6	3.0	0	17
Income-to-poverty ratio[a]	2.28	1.47	0.00	8.47

Source: US DHHS 1997.
Note: NHANES III = Third US National Health and Nutrition Examination Survey, 1988–1994.
Unweighted N = 9,813. Statistics weighted to national level using sampling weights provided with the NHANES (US DHHS 1997).
[a] Income-to-poverty ratio = family income divided by the Federal Poverty Threshold for a family of that size and age composition.

Table 6.5. Comparison of sample with target population

Birth weight, sociodemographic characteristics, and smoking behavior, NHANES III sample, 1988–1994, and all US births, 1997

	NHANES III sample[abc]	All US births, 1997[d]
Birth weight		
Median (grams)	3,402	3,350
% Low birth weight	6.8	7.5
(<2,500 grams)		
Sociodemographic factors		
Race/ethnicity		
Non-Hispanic white	73.4	68.4[e]
Non-Hispanic black	16.9	17.0
Mexican American	9.7	14.6
Mother's age		
Median (years)	26.0	26.7
% Teen mother	12.5	12.7
Mother's education		
Median (years)	12.0	12.8
% <High school	21.6	22.1
% = High school	35.0	32.4
Mother smoked while	24.5	13.2
pregnant (%)		
Number of cases	9,813	3,880,894

NHANES III = Third US National Health and Nutrition Examination Survey, 1988–1994.
[a] Weighted to population level using weights provided with the NHANES III (Westat 1996); sample size is unweighted.
[b] Information for NHANES III is calculated from data extracted from National Center for Health Statistics (US DHHS 1997).
[c] Includes non-Hispanic white, non-Hispanic black, and Mexican American infants with complete information on family income, birth weight, maternal age, and education.
[d] Information for all US births is from Ventura et al. (1999) except median mother's age (Mathews and Hamilton 2002).
[e] For consistency with the NHANES III sample, racial composition of US births is reported as a percentage of births that are non-Hispanic white, non-Hispanic black, or Mexican American, excluding births of other Hispanic origins or racial groups. When all racial/ethnic groups are considered, the racial composition is 60.1% non-Hispanic white, 15.0% non-Hispanic black, 12.9% Mexican American, 5.4% other Hispanic origin, and 6.6% other racial groups.

of the NHANES III survey sample used to study birth weight are compared against those among all births that occurred nationally at about the same time. This type of comparative table could also present information on alternative measures of a concept, such as ratings of items from separate samples at different points in time or from each of several sources (not shown).

Bivariate Tables
Bivariate, or two-way, tables show the relationship between two variables. Common types of bivariate tables are cross-tabulations, those that present differences in means or other statistics for one variable according to values of a second variable, and correlations. The nature of your variables—categorical or continuous—will determine which type of table applies to your topic.

CROSS-TABULATIONS
A cross-tabulation shows the joint distribution of two categorical variables—how the overall sample is divided among all possible combinations of those two variables. Table 6.6 shows how poverty rates differ according to age group, calculated from a cross-tabulation of two variables: age group and poverty status. Readers could calculate the number of persons in each poverty category from the total number of persons and the percentage poor in each age group[2] (see "Which Numbers to Include" below).

DIFFERENCES IN MEANS
Bivariate tables are also used to present statistics for one or more continuous variables according to some categorical variable. Table 6.2

Table 6.6. Bivariate table: Rates of occurrence based on a cross-tabulation
Poverty rates (%) by age group, United States, 2012

	Age group (years)			
	<18	18–64	65+	All ages
Population (1,000s)	73,719	193,642	43,287	310,648
% Poor	21.8	13.7	9.1	15.0

Source: DeNavas-Walt, Proctor, and Smith 2013.

shows how AIDS knowledge varies by language group, presenting mean scores for two topic areas ("likely" and "unlikely" modes of AIDS transmission) for each of three language groups.

CORRELATIONS

A bivariate table can present correlations among continuous variables. In table 6.7, each interior cell holds the pairwise correlation coefficient between the variables named in the associated row and column. For instance, in the late 1990s, the correlation between the child poverty rate and the unemployment rate in New Jersey's 21 counties was 0.75. Notes to the table show calculations, cite sources to define potentially unfamiliar measures used in the table, and define symbols denoting statistical significance level.

To present more detailed information about the joint distribution of two continuous variables, use a line graph or scatter chart (see chapter 7).

Three-Way Tables

Three-way tables present information on associations among three variables or sets of related variables, such as the joint distribution of three categorical variables. One way to show a three-way relationship is to use column spanners. In table 6.8, the columns contain two variables—gender and type of drug—with rows for selected major cities in the United States. The spanners divide the table into sections for males and females. Within each of those spanners is a column for each of three types of drugs and for all drugs combined. This structure facilitates comparison across types of drugs within each gender because the drugs are in adjacent columns. If you want to emphasize comparison of drug use for males against that for females, place type of drug in the upper column spanner (there would be four such spanners) with columns for each gender arranged underneath.

This type of design works only if the two variables used in the column spanners and the columns below have no more than a few categories apiece. For variables with more categories, use panels or a chart to present three-way relationships.

Table 6.7. Bivariate table: Pairwise correlations among continuous variables

Correlations among county-level demographic and economic characteristics, New Jersey, 1999–2001

	Demographic characteristics					Economic characteristics		
Characteristics	Black pop.	Black ID[a]	Pop. density	Total pop.	Non-Engl. speakers (%)	Unemp. rate (%)	Income inequality[b]	Child pov. rate (%)
Demographic								
Black population	1.00							
Black ID[a]	0.15	1.00						
Population density (persons/mi.²)	0.39	0.12	1.00					
Total population	0.31	0.12	0.58[d]	1.00				
Non-English speakers (%)	0.39	0.05	0.91[d]	0.49[c]	1.00			
Economic								
Unemployment rate (%)	0.31	−0.17	0.22	−0.16	0.33	1.00		
Income inequality[b]	0.10	0.94[d]	0.10	0.09	−0.02	−0.15	1.00	
Child poverty rate (%)	0.69[d]	−0.07	0.53[c]	0.12	0.61[d]	0.75[d]	−0.06	1.00

Source: Quality Resource Systems 2001.

Note: *N* = 21 counties.

[a] ID = Index of dissimilarity, a measure of residential segregation; see James and Taeuber (1985).

[b] Gini coefficient, a measure of income inequality; see Levy (1987).

[c] $p < 0.05$.

[d] $p < 0.01$.

Table 6.8. Three-way table with nested column spanners

Drug use by arrestees in selected major US cities by type of drug and sex, 1999

| City | Percentage testing positive[a] | | | | | | | | | |
| | Male | | | | Female | | | | |
	Any drug[b]	Marijuana	Cocaine	Heroin	Any drug	Marijuana	Cocaine	Heroin
Atlanta, GA	76.7	44.4	51.3	4.3	77.2	33.5	62.0	4.5
Chicago, IL	74.4	44.6	41.7	20.1	76.9	26.5	64.3	32.4
Los Angeles, CA	62.4	32.3	35.6	5.5	61.6	21.0	36.7	8.2
Miami, FL	66.0	36.2	49.2	3.4	NA[c]	NA	NA	NA
New York, NY	74.7	40.8	44.2	15.2	81.3	26.2	65.1	21.1
Washington, DC	68.9	34.9	37.7	16.0	NA	NA	NA	NA

Source: US Census Bureau 2002b.

[a] Based on data from the Arrestee Drug Abuse Monitoring Program: US National Institute of Justice, 2000 Arrestee Drug Abuse Monitoring: Annual Report, available at http://www.ncjrs.org/txtfiles1/nij/193013.txt.

[b] Includes other drugs not shown separately.

[c] NA = not available.

ORGANIZING TABLES TO COORDINATE WITH YOUR WRITING

As you write about the patterns shown in your tables, proceed systematically, comparing numbers either across the columns or down the rows of your table. To describe both types of patterns, create separate paragraphs for the "down the rows" and "across the columns" comparisons (see appendix A). Decide on the main point you want to make about the data using one or more of the principles described below and in Miller (2007a). Arrange the rows and columns accordingly, then describe the numbers in the same order they appear in the table. If possible, use the same organizing principles in all the tables within a document, such as tables reporting descriptive and inferential statistics for the same set of variables. See "Organizing Charts to Coordinate with Your Writing" in chapter 7 for an illustration of the advantages and disadvantages of the different organizing principles.

When reporting results for ordinal variables, the sequence of items in rows or columns will be obvious. List the possible response categories in their ranked order: from "Excellent" to "Poor," or from "Agree strongly" to "Disagree strongly," for example, either in ascending or descending order, depending on how you prefer to discuss them. Arrange information for each of several dates in chronological order.

For nominal variables (such as religion or race) or for tables that encompass several different variables (such as AIDS knowledge topics), the categories or variables do not have an inherent order. In those instances, use one or more of the following principles to organize them.

Theoretical Grouping

Arranging items into theoretically related sets can be very effective. Using panels to separate likely and unlikely modes of AIDS transmission in table 6.2 reveals important distinctions between the items that would be obscured if they were listed in one undifferentiated block. The accompanying discussion can then emphasize that the former topics were much better understood than the latter without asking readers to zigzag through the table to find the pertinent numbers.

Most analyses concern relationships among two or three variables, with other variables playing a less important role. Logically, you will discuss the key variables first, so put them at the top of your table. For instance, a table of univariate descriptive statistics for a study of how race and socioeconomic characteristics relate to birth weight might put the dependent variable (birth weight) in the top row, followed by the main independent variable or variables (race and socioeconomic factors), and then other factors considered in the study (e.g., smoking behavior; see table 6.5).

Empirical Ordering

For many tables presenting distributions or associations, an important objective is to show which items have the highest and the lowest values and where other items fall relative to those extremes. If this is your main point, organize your univariate or bivariate table in ascending or descending order of numeric values or frequency, as with the categories of race/ethnicity in table 6.3.[3] An international health report might list countries in rank order of infant mortality, life expectancy, or death rates from AIDS, for example. If the ranking varies for different variables shown in the table, decide which you will emphasize, then use it as the basis for organizing the rows.

Alphabetical Ordering

Few substantively meaningful patterns happen to occur alphabetically, hence alphabetical order is usually a poor principle for arranging items within tables to be discussed in the text. On the other hand, alphabetical order is often the best way to organize data in appendix tables or other large data tabulations that are not accompanied by written guidance. In such cases, using a familiar convention helps readers find specific information quickly. The daily stock market report of opening, closing, high, and low prices of thousands of stocks is a well-known example.

Order of Items from a Questionnaire

Unless your analysis is mainly concerned with evaluating the effects of questionnaire design on response patterns, do not list items in the order they appeared on the questionnaire. Again, this order is unlikely to correspond to underlying empirical or theoretical patterns, so your table and description will not match.

Multiple Criteria for Organizing Tables

For tables with more than a few rows of data, a combination of approaches may be useful for organizing univariate or bivariate statistics. You might group items according to conceptual criteria, then arrange them *within* those groups in order of descending frequency or other empirical consideration. In table 6.2, knowledge of AIDS transmission is first grouped into likely and unlikely modes of transmission, then in descending order of knowledge within each of those classifications.

Sometimes it makes sense to apply the same criterion sequentially, such as identifying major theoretical groupings and then minor topic groupings within them. Political opinion topics could be classified into domestic and foreign policy, for example, each with a major row heading. Within domestic policy would be several items apiece on education, environment, health, transportation, and so forth, yielding corresponding subcategories and sets of rows. Foreign policy would also encompass several topics.

If you have organized your table into several theoretically or empirically similar groups of items, alphabetical order can be a logical way to sequence items *within* those groups. For example, data on the 50 United States are often grouped by major census region, then presented in alphabetical order within each region. Alphabetical order within conceptual or empirical groupings also works well if several items have the same value of the statistics reported in the table (e.g., mean or frequency).

Conventions about placement of "total" rows vary, with some publications placing them at the top of the table or panel, others at the bottom. Consult your publisher's instructions to decide where to place your total row.

TECHNICAL CONSIDERATIONS

Which Numbers to Include

A table is a tool for presenting numeric evidence, not a database for storing data or a spreadsheet for doing calculations. Except for a data appendix or a laboratory report, omit the raw data: readers don't need to wade through values of every variable for every case in a large data set. Generally you will also leave out numbers that represent intermediate steps in calculating your final statistics. Like a carpenter, do the messy work (data collection and calculations) correctly, then present a clean, polished final product. Decide which numbers are needed to make the table's substantive points, then keep the other data out of the table, where your readers don't trip over it on their way to the important stuff.

The output from a statistical program is usually a poor prototype of a table for publication: such output often includes information that isn't directly relevant to your research question and thus should be omitted from your table. For example, output from a cross-tabulation usually shows the count (number of cases), row percentage, column percentage, and percentage of the overall (grand) total for every cell. Determine which of those statistics answer the question at hand— usually *one* of the percentages and possibly the number of cases for each cell—and report only those numbers in your table. In most instances, report the number of cases only for the margins of a cross-tabulation (as in table 6.6), because the counts for interior cells can be calculated from the marginals (the row and column subtotals found at the edges, or margins, of the cross-tabulation) and the interior percentages.

Cross-tabulations of dichotomous (two-category) variables in particular include a lot of redundant information. If you report the percentage of cases that are US born, the percentage foreign born is unnecessary because by definition, it is 100% minus the percentage US born. Likewise for variables coded true/false, those that indicate whether some event (e.g., a divorce, cancer diagnosis) did or did not happen, whether a threshold was or wasn't crossed (e.g., low birth weight, the legal alcohol limit), or other variants of yes/no.[4] Hence a tabulation of approval of legal abortion (a yes/no variable) by a six-category religion variable will yield 12 interior cells, eight marginals, and the grand total, each of which contains counts and one or more percentages. In your final table, only seven of those numbers are needed to compare abortion attitudes across groups: the share of each religious group that believes abortion should be legal, and the approval rate for all religions combined; subgroup sample sizes could also be included if they aren't reported elsewhere.

Number of Decimal Places and Scale of Numbers

Within each column, use a consistent scale and number of decimal places. For instance, do not switch from grams in one row to kilograms in other rows. Likewise, keep the scale and number of decimal places the same for all columns reporting numbers measured in similar units: if all your columns show death rates, use a uniform scale (e.g., deaths per 100,000 persons across the board, not per 1,000 in some columns).

Follow the guidelines in chapter 4 regarding the following:

- Number of digits and decimal places that inform but do not overwhelm. Change the scale or use scientific notation to avoid presenting overly long numbers or those with many zeros as placeholders.
- Conventions about decimal places for certain kinds of numbers: two decimal places for small monetary denominations, none for integers
- Precision of measurement that is consistent with the original data collection
- Sufficient detail to evaluate statistical test results (e.g., for p-values or test statistics)

Alignment

There are certain standard or sensible ways to align the contents of different table components:

- Left-justify row labels, then use indenting to show subgroups.
- Use decimal alignment in the interior cells to line up the numbers properly within each column, especially if symbols are used in some but not all rows (e.g., to denote statistical significance). Right alignment works too, assuming you have used a consistent number of decimal places for all numbers in the column and no symbols are needed.
- Center column titles over the pertinent column, and column spanners over the range of columns to which they apply.

Portrait versus Landscape Layout

Tables can be laid out in one of two ways: *portrait* (with the long dimension of the page vertical, like table 6.3), or *landscape* (with the long dimension horizontal, like table 6.1). For print documents and Web pages, start with a portrait layout because the accompanying text pages are usually vertical. For slides or chartbooks, start with a landscape layout to match the rest of the document.

These general considerations aside, pick a layout that will accommodate the number of rows and columns needed to hold your information. If you have more than a dozen rows, use a portrait layout or create a multipanel landscape table that will flow onto more than one page. Unless your column labels and the numbers in the corresponding interior cells are very narrow, four to five columns are the most that can fit in a portrait layout, up to 12 narrow columns in a landscape layout.

Consider alternative arrangements of variables in the rows and columns. If you are cross-tabulating two variables, there is no law that decrees which variable must go in the rows. Take, for example, a table comparing characteristics of geographic entities: the 50 United States are virtually always listed in the rows because a 50-column table would be ungainly. On the other hand, the six populated continents easily fit within the columns of a landscape table. Which variable to put in the rows is determined by the number of categories (countries or continents), not the concept being measured.

Type Size

For your tables, use a type size consistent with that in your text—no more than one or two points smaller. With the possible exception of reference tables, tiny scrunched labels with lots of abbreviations and microscopically printed numbers are usually a sign that you are trying to put too much into one table. Redesign it into several tables, each of which encompasses a conceptually related subset of the original table. Arrange a large appendix table into a few panels per page using one or more of the criteria explained above to divide and organize the variables so readers can find information of interest easily.

Table and Cell Borders

Many word processing programs initially create tables with grid-lines delineating the borders between cells. However, once you have typed in the row and column labels, most of those lines are no longer needed to guide readers through your table. Places where it is useful to retain lines within a table include borders between panels, and lines to designate a column spanner.

Word processing programs offer table templates or auto format-ting—predesigned formats complete with fancy colors, fonts, shad-ing, and lines of different thickness. While some lines and other features can make it easier to read a table, others simply add what Edward Tufte (2001) refers to as "nondata ink": aspects of the design that distract readers rather than adding to the function of the table. Design your tables to emphasize the substantive questions and perti-nent data, not superfluous eye candy.

Formatting for Different Publications

Table formatting varies by discipline and publisher. Some require titles to be left-justified, others centered. Some require all capital let-ters, others mixed upper- and lowercase. Many journals have specific guidelines for labeling footnotes to tables and using other symbols within tables. Requirements for punctuation and use of lines within the table also vary. Consult a manual of style for your intended pub-lisher before you design your tables. Even if you aren't required to fol-low specific guidelines, be consistent as you format your tables: do not left-justify one table title, then center the title for the next table, or label footnotes to one table with letters but use symbols to denote footnotes in a second.

DRAFTING YOUR TABLES

Conceptualize the contents and layout of a table early in the writing process, certainly before you start typing the information into a word processor and possibly even before you collect or analyze the data. Identifying the main question to be addressed by each table helps you anticipate the statistics to be calculated. Thinking ahead about the specific points you will make about patterns in the table helps you design a layout that coordinates with the description.

Drafting Your Table with Pencil and Paper

To create an effective table, plan before you type. Separating these steps helps you think about a layout and labels that emphasize the substantive concepts to be conveyed before you get caught up in the point-and-click task of creating the table on a computer. By not planning ahead, you are more likely to write incomplete titles or labels, create too few columns or rows, overlook important features like column spanners or footnotes, and to arrange these elements poorly. You must then go back and reconstruct the table on the screen—quite a hassle if the numbers have already been typed in—increasing the risk that numbers end up in the wrong cells of the table.

To test possible layouts for your tables, use scrap paper, a pencil, and an eraser. Don't skimp on paper by trying to squeeze drafts of all four (or however many) tables you need to plan onto one page. Use a full page for each table. Expect to have to start over a couple of times, especially if you are new to planning tables or are working with unfamiliar concepts, variables, or types of statistical analyses.

DETERMINING THE SHAPE AND SIZE OF YOUR TABLE

Create one column for each set of numbers to be displayed vertically, then add a column for row labels. Make the column for row labels wide enough to accommodate a short, intelligible phrase that identifies the contents (and sometimes units) in each row. Most numeric columns can be narrower (see below).

Count how many variables you will be displaying in the rows, then add rows to accommodate the table title and column headings. Depending on the content and organization of your table, you may need additional rows to fit column spanners, labels for panels, information on the overall sample (e.g., total sample size), results of statistical tests, or simply white space to increase ease of reading.

Once you know how many rows and columns you need, assess whether a landscape or portrait layout will work better. For tables with approximately equal numbers of rows and columns, try it both ways to see which fits the information better and is easier to read. Orient your scrap paper accordingly, then draw in gridlines to delineate the rows and columns. Erase gridlines to show column spanners and draw in horizontal lines to differentiate panels. The idea is to have a grid within which you can test out the labels and other components of the table. You will probably redesign and redraw it a couple of times before you are satisfied, so don't bother to measure exact spacing or draw perfect, straight lines in your rough drafts. The software will do that for you, once you have decided what the table should look like.

INTERIOR CELLS

The interior cells of the table are where your numbers will live. When planning column widths, consider the following questions:

- What is the maximum number of digits you will need in each column?
- Do you need to include unit indicators (e.g., $, %), thousands' separators (e.g., the comma in "1,000"), or other characters that will widen your column?
- Will you be using symbols within the table cells, to key them to a footnote or indicate statistical significance, for example?

Evaluating Your Table Layout

Before you type your table into a word processor, evaluate it for completeness and ease of understanding. To test whether your table can stand alone, pick several cells within the table and see whether you can write a complete sentence describing the identity and meaning of those numbers using only the information provided in the table. Better yet, have someone unfamiliar with your project do so.

Creating Your Table in a Word Processor

Word processing software can be a real boon for creating a table: simply tell your computer to make a seven-column by 12-row table on a landscape page and voila! However, some word processors think they know what you want better than you do, and will automatically format aspects of your table such as page layout, alignment, type size, and whether text wraps to the next line. After you have created the basic table structure on the computer, save it, then check carefully that each part of the table appears as you showed it on your rough draft. Occasionally, the computer's ideas will improve upon yours, however, so consider them as well.

A few hints:

- *Before* you begin to create the table, inform the word processor whether you want a portrait or landscape page layout, then save the document. Then when you specify the desired number of columns, their width will be calculated based on the full available width (usually 6.5″ for portrait, 9.5″ for landscape, assuming 1″ margins all around). Much easier than manually resizing columns from portrait to landscape after the fact ...
- Specify alignment for each part of the table after creating the initial grid. An alternative in some software programs is to impose a selected, preformatted design for your table (see your software manual or Help menu for more information).
- Alter appearance of table and cell borders.

    Once titles and labels are in place, omit many cell borders for a cleaner look.

    Resize column widths to accommodate row labels and numbers.

    Delineate panels within the table. If you have omitted most row borders, use a single, thin line or a blank row to separate panels; if you have retained borders between all rows, use a thicker line, a double line, or a blank row between panels.

As you transfer your design into the word processor, you may discover that a different layout will work better, so learn from what appears on the screen and then revise it to suit.

CHECKLIST FOR CREATING EFFECTIVE TABLES

- Title: write an individualized title for each table.

    State the purpose or topic of that table.

    Include the context of the data (the W's).

    Identify units, if the same for all or most variables in the table.

- Label each row and column.

    Briefly identify its contents with a short phrase, *not* acronym or numeric code.

    Specify units or categories if not summarized in table title.
- Footnotes

    Identify the data source (if not in table title).

    Define all abbreviations and symbols used within the table.
- Structure and organization

    Use indenting or column spanners to show how adjacent rows or columns relate.

    Apply theoretical and empirical principles to organize rows and columns.

    - For text tables, coordinate row and column sequence with order of discussion.
    - For appendix tables, use alphabetical order or another widely known principle for the topic so tables are self-guiding.

    Report the fewest number of digits and decimal places needed for your topic, data, and types of statistics.

    Use consistent formatting, alignment, and symbols in all tables in a document.
- Check that the table can be understood without reference to the text.

CREATING EFFECTIVE CHARTS

Well-conceived charts provide a good general sense of pattern, complementing a prose description by graphically depicting the shape and size of differences between numbers. The many uses of charts in paper, speeches, posters, or briefs about quantitative analyses include the following:

- Displaying sample composition in terms of the key variables in the analysis
- Portraying bivariate or three-way associations among variables
- Facilitating visual hypothesis-testing with the addition of confidence intervals around point estimates
- Showing the sensitivity of results to alternative assumptions or definitions

Although these benefits are most pronounced for speeches, short documents, and products for lay audiences, they are often underutilized in scientific papers. In any of those contexts, charts are an effective, efficient way to convey patterns, keeping the focus on your story line rather than requiring readers to do a lot of mental gymnastics to calculate or compare numbers.

I begin this chapter with a quick tour of the anatomy of a chart, followed by general guidelines about features shared by several types of charts. Using examples of pie, bar, line, and scatter charts and their variants, I then show how to choose the best type of chart for different topics and types of variables. I illustrate the importance of using theoretical or empirical criteria to organize charts to coordinate with your writing, and finally discuss additional design considerations and common pitfalls in chart creation. For other resources on chart design, see Tufte (1990, 1997, 2001), Briscoe (1996), and Zelazny (2001).

ANATOMY OF A CHART

Many of the principles for designing effective tables apply equally to charts.

- Label the chart so readers can identify its purpose and interpret the data from the titles and labels alone.
- Make the chart self-contained, including context (W's), units, category names, data sources, and definitions of abbreviations and symbols.
- Design each chart to promote understanding of the patterns in that chart and to coordinate with the written description.
- Create charts that emphasize the evidence related to the research question rather than drawing undue attention to the structure of the charts themselves.

Chart Titles

The same principles that guide creation of table titles also work for charts.

- Specify the topic and W's in each chart title. A short restatement of the research question or relationships shown in the chart often works well.
- Use the title to differentiate the topic of each chart from those of other charts and tables in the same document.

Axis titles, labels, and legends identify the concepts and units or categories of the variables in charts, much as row and column labels do in tables.

Axis Titles and Axis Labels

Charts that illustrate the relations between two or more variables usually have an *x* (horizontal) axis and a *y* (vertical) axis. Give each axis a title that identifies its contents and units of measurement, and include labels for categories or values along that axis. Write brief but

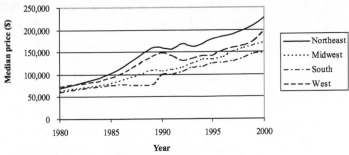

Median sales price of new single-family homes, by region, United States, 1980–2000

Figure 7.1. Anatomy of a chart: Multiple-line chart
Source: US Census Bureau 2001a

informative axis titles and labels, using short phrases or single words instead of acronyms whenever possible. In the axis *title* name the overall concept ("Year" for the *x* axis title in figure 7.1), then assign axis *labels* to identify values (1980, 1985, . . . 2000; figure 7.1) or category names (Saudi Arabia, Russia, etc.; figure 7.4) along the axis.

For continuous variables, minimize clutter by marking major increments of the units, aiming for 5 to 10 value labels on each axis. Remember, charts are best used when precise values aren't important, so your axis labels need only show approximate values. To choose an appropriate increment for the axis scale, consider the range, scale, precision of measurement, and the units or categories in which the data were collected or classified. To present census data for 1960 through 2010, for example, make those dates the limits of the axis values, and label 10-year increments along the *x* axis to match the data collection years.

In the chart title, name the general concepts or variables ("median sales price" in figure 7.1), then give specific units for that dimension ($) in the axis title. For pie charts, which don't have axes, identify units in the title, footnote, or data labels.

Legends

Use a legend to identify the series or categories of variables that are not labeled elsewhere in the chart. In figure 7.1, the legend specifies which line style corresponds to each of the four regions. See below for how to use legends in other types of charts.

Data Labels

Data labels are typed numeric values adjacent to the pertinent slice, point, or bar in a chart (e.g., the reported percentages in figure 7.2b). To keep charts simple and readable, use data labels sparingly. Again, the main advantage of a chart is that it can illustrate general levels or patterns, which will be evident without data labels if your chart has adequate titles. Complement the general depiction in the chart with your text description, reporting exact values of selected numbers to document the patterns (see "Generalization, Example, Exception" in chapter 2, and appendix A for guidelines). If your audience requires exact values for all numbers in the chart, replace the chart with a table or include an appendix table rather than putting data labels on every point.

Reserve data labels for reference points or for reporting the value associated with a pie or stacked bar chart, such as total number of cases or total value of the contents of the pie or stacked bar (e.g., total dollar value of annual outlays in figures 7.2 and 7.7b). See pertinent sections below for more on data labels.

CHART TYPES AND THEIR FEATURES

Charts to Illustrate Univariate Distributions

Univariate charts present data for only one variable apiece, showing how cases are distributed across categories (for nominal or ordinal variables) or numeric values (for interval or ratio variables).

PIE CHARTS

Most people are familiar with pie charts from elementary school. A pie chart is a circle divided into slices like a pizza, with each slice representing a different category of a variable, such as expenditure categories in the federal budget (figure 7.2). The size of each slice illustrates the relative size or frequency of the corresponding category. Identify each slice either in a legend (figures 7.2a and b) or in a value label adjacent to each slice (figure 7.2c). Although you can also label each slice with the amount or percentage of the whole that it contributes (figure 7.2b), the basic story in the chart is often adequately illustrated without reporting specific numeric values: is one slice much larger (or smaller) than the others, or are they all about equal? Pie

a. US federal outlays by function, 2000

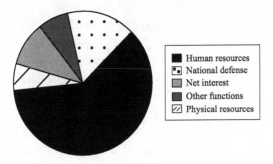

Total outlays: $1.8 trillion

b. US federal outlays by function, 2000

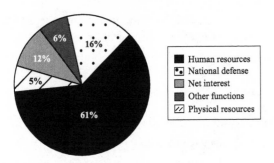

Total outlays: $1.8 trillion

c. US federal outlays by function, 2000

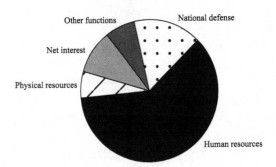

Total outlays: $1.8 trillion

Figure 7.2. Three variants of pie charts to illustrate composition: (a) Without data labels, (b) with data (numeric value) labels, (c) with value labels
Source: US Office of Management and Budget 2002.

charts also work well to display the composition of a study sample in terms of a nominal characteristic. Create one pie to illustrate each trait (e.g., one for race composition, another for gender).

To compare two or three pie charts that differ in the total quantity they represent, you can make them proportionate to their respective totals. For instance, if the total budget in one year was twice as large as in another, create the former pie with twice the area of the latter.

Use pie charts to present composition or distribution—how the parts add up to the whole. Each pie chart shows distribution of a single variable, such as racial composition of the study sample. Because they illustrate composition, pie charts can be used only for single-response variables whose values are mutually exclusive (chapter 4)—after all, the slices of a pizza don't overlap one another in the pan.

- Display only one variable per pie chart: either age or gender distribution—not both.[1]
- Don't use pie charts to compare averages or rates across groups or time periods. Those dimensions don't have to add up to any specifiable total, so a pie chart, which shows composition, doesn't fit the topic. E.g.,
  - High school graduation rates for boys and for girls *don't* add up to the graduation rate for the two genders combined.
  - Average temperatures in each of the 12 months of the year are not summed to obtain the average temperature for the year. Such logic would yield an average annual temperature of 664°F for the New York City area.
- Don't use a pie chart to contrast measures of quantitative comparison such as rates, ratios, percentage change, or average values of some outcome, which are not constrained to total 100; instead, use a bar or line chart.
- Don't use a pie chart to present multiple-response variables, such as which reasons influenced buyers' choice of a car model; those responses are not mutually exclusive. Instead, use a bar chart to show the frequency of each response.
- Avoid pies with many skinny slices. Consider combining rare categories unless one or more of them are of particular interest to your research question. Or create one pie that includes the 3

or 4 single most common responses with a summary slice for "other," then a second pie that shows a detailed breakdown of values within the "other" category.

In addition to the above pie-chart no-nos, some aspects of composition are more effectively conveyed with a different type of chart.

- Use a histogram to present the distribution of values of an ordinal variable, especially one with many categories. Histograms show the order and relative frequency of those values more clearly than a pie.
- To compare composition across more than three groups (e.g., distribution of educational attainment in each of 10 countries), use stacked bar charts, which are easier to align and compare than several pie charts. Or create a multipanel histogram.

HISTOGRAMS

Histograms are a form of simple bar chart used to show distribution of variables with values that can be ranked along the x axis. Use them to present distribution of an ordinal variable, such as the share of adults that fall into each of several age groups (figure 7.3), or an interval (continuous) variable with 20 or fewer values. For a ratio variable or interval variable with more than 20 values, such as IQ score, use a line chart to present distribution.

Array the values of the variable across the x axis and create a bar to show the frequency of occurrence of each value, either number of cases or percentage of total cases, measured on the y axis. To accurately portray distribution of a continuous variable, don't leave horizontal space between bars for adjacent x values. Bars in a histogram should touch one another (as in figure 7.3) unless there are intervening x values for which there are no cases. For example, in figure 4.3d, no cases have any of the values 3 through 9, so there is a gap above the labels for those values, between the bars showing frequency of the x values 2 and 10. Alternatively, use a line chart or box-and-whisker plot to illustrate distribution of a single variable (see respective sections below).

Don't use histograms to display distribution of nominal variables

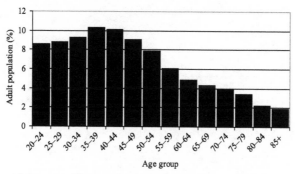

Age distribution of US adults, 2000

Figure 7.3. Histogram to illustrate distribution of an ordinal variable
Source: US Census Bureau 2002d

such as religion or race, because their values do not have an inherent
order in which to arrange them on the x axis. Instead, use a pie chart.

A histogram can be used to display distribution of an ordinal variable with unequal width categories; see cautions under "Line Chart
for Unequally Spaced Ordinal Categories" in the section on "Common
Errors in Chart Creation" below.

Charts to Present Relationships among Variables

BAR CHARTS

Simple bar chart

A simple bar chart illustrates the relationship between two variables—a categorical predictor variable on the x axis, and a continuous
outcome variable on the y axis. They can also be used to report the percentage occurrence of one category of a dichotomous (two-category)
variable, such as the percentage of infants that are low birth weight
for each of several racial groups. Most dimensions of quantitative
comparison—value, difference, ratio, or percentage change—can be
shown in a bar chart, making it an effective tool for comparing values
of one variable across groups defined by a second, categorical variable.
Create one bar for each group with the height of the bar indicating the
value of the outcome variable for that group, such as mean daily crude
oil production (y axis) for each of five countries (x axis, figure 7.4).

To format a simple bar chart, place the continuous variable on the
y axis and label with its units. The variable on the x axis is usually

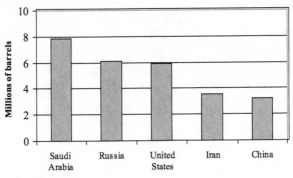

Mean daily crude oil production in the five leading countries, 1999

Figure 7.4. Simple bar chart
Source: US Census Bureau 2002d

nominal or ordinal. Arrange the categories of nominal variables such as race or country of residence in meaningful sequence, using theoretical or empirical criteria. Display the categories of ordinal variables or values of a continuous variable in their logical order (e.g., income group, letter grades). Simple bar charts don't need a legend because the two variables being compared are defined by the axis titles and labels; hence the same color is used for all the bars.

Clustered bar chart

Use a clustered bar chart to introduce a third variable to a simple bar chart, illustrating relationships among three variables—a continuous outcome variable by two categorical predictors. For example, figure 7.5a shows how emergency room (ER) use for asthma varies according to race and income level simultaneously. One predictor (income group) is shown on the x axis, the other predictor (race) in the legend, and the outcome (odds ratio of ER use for asthma) on the y axis.

Clustered bar charts also work well to show patterns for multiple-response items (e.g., how votes for the top three candidates for school board varied according to voter's party affiliation) or a series of related questions. For example, figure 7.6 includes one cluster for each of 10 AIDS knowledge topics, with a bar for each language group. The height of each bar shows the percentage of the pertinent language group that answered that question correctly.

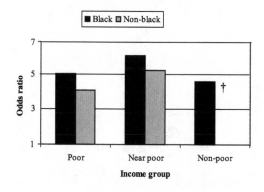

a. Odds ratios of emergency room visits for asthma, by race and income, United States, 1991

b. Odds ratios of emergency room visits for asthma, by race and income, United States, 1991

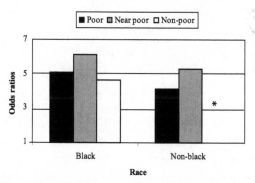

Figure 7.5. Two versions of clustered bar chart: Patterns by two nominal variables (a) by income group, and (b) by race
Source: Miller 2000b; data are from US DHHS 1991.
Compared to non-black non-poor.
* Difference across income groups significant at $p < 0.05$ for non-blacks only.
† Difference across racial groups within income group significant at $p < 0.05$ for non-poor only.

Finally, clustered bar charts can be used to show distribution of one variable for each of two or three groups, with the histogram for each group comprising one cluster within the chart.

To format a clustered bar chart, place one categorical predictor on the *x* axis, the other categorical predictor in the legend, and the

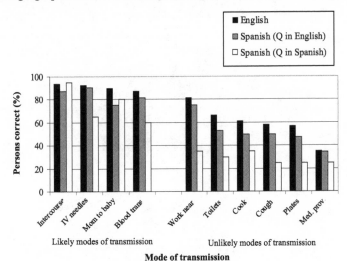

Knowledge of AIDS transmission modes by topic, language spoken at home, and language of questionnaire (Q), New Jersey, 1998

■ English
▨ Spanish (Q in English)
□ Spanish (Q in Spanish)

Likely modes of transmission   Unlikely modes of transmission

Mode of transmission

Figure 7.6. Clustered bar chart: Series of related outcomes by a nominal variable
Source: Miller 2000a
Note: See table 6.2 for detailed wording of questions.

continuous outcome variable on the *y* axis. Label each cluster on the *x* axis for the corresponding category of the first predictor variable, then include a legend to identify values of the second predictor.

To decide which predictor variable to show on the *x* axis and which to put in the legend, anticipate which contrast you want to emphasize in your description. Although the description of either version of figure 7.5 will include both the income pattern of ER use and the racial pattern of ER use (see appendix A), each arrangement highlights a different contrast: Figure 7.5a underscores the pattern of emergency room use for asthma across income groups (on the *x* axis). Figure 7.5b presents the same data but reverses the variables in the legend (income group) and the *x* axis (race), highlighting the comparisons across racial groups.

Stacked bar chart
Create a stacked bar chart to show how the distribution of a multi-category variable differs according to another characteristic, such as

how age composition varies by race, or how different causes of death add up to the respective overall death rates for males and for females. Because they illustrate the contribution of several parts to a whole (composition), stacked bar charts can be used only for variables with mutually exclusive categories, just like pie charts. For multiple-response items, use a clustered bar chart.

There are two major variants of stacked bar charts: those that show variation in level and those that show only composition of each group's whole. To emphasize differences in level while also presenting composition, construct a stacked bar chart that allows the height of the bar to reflect the level of the outcome variable. Figure 7.7a shows how total federal outlays (y axis) were divided among major functions with data at 10-year intervals (x axis). For each year, the dollar value of outlays in each category is conveyed by the thickness of the respective slice (defined in the legend), and the value of total outlays for all functions combined by the overall height of the stack.

If there is wide variation in the level of the outcome variable, however, this type of stacked bar chart can obscure important intergroup differences in the distribution of those components. For example, with more than a 40-fold increase in total federal outlays between 1950 and 2000, it is virtually impossible to assess the relative contribution of each category in the early years based on figure 7.7a.

To compare composition when there is more than a three-fold difference between the lowest and highest y values across groups or periods, create a stacked bar chart with bars of equal height, and show percentage distribution in the stacked bar. This variant of a stacked bar chart highlights differences in composition rather than level. Figure 7.7b shows that the share of outlays for defense dropped from roughly 50% in 1960 to 16% in 2000. Because this version of a stacked bar chart does not present information on level (e.g., outlays in billions of dollars) on the y axis, report that information in a separate table, at the top of each stack (as in figure 7.7b), or in a footnote to the chart.

In formatting stacked bar charts, the variables on the x axis and in the slices (legend) must be categorical; the variable on the y axis is continuous. On the x axis, put the variable that is first in the causal chain, then show how the legend variable differs within those categories. For stacked bar charts in which the height of the bar reflects the level

a. US federal outlays by function, 1950–2000

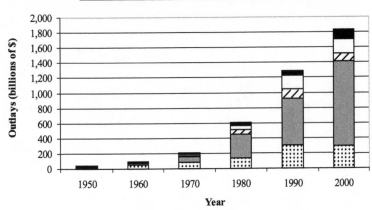

b. US federal outlays by function, 1950–2000

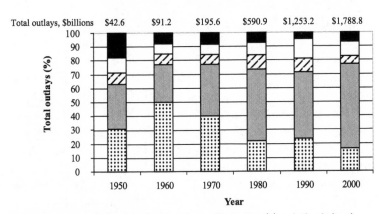

Figure 7.7. Two versions of stacked bar charts, illustrating (a) variation in level, (b) percentage distribution

Source: US Office of Management and Budget 2002.

Note: Human resources include education, training, employment, and social services; health; Medicare; income security; Social Security; and veterans benefits and services. Physical resources include energy; natural resources and environment; commerce and housing credit; transportation; and community and regional development. Other functions include international affairs; general science, space, and technology; agriculture; administration of justice; general government; and allowances

of the outcome variable, the units on the y axis are those in which the variable was originally measured. In figure 7.7a, for example, the y axis shows federal outlays in billions of dollars. For stacked bar charts that emphasize composition (e.g., figure 7.7b), the y axis units are percentage of the overall value for that stack, and by definition, the height of all the bars is the same since each bar reflects 100% of that year's outlays.

LINE CHARTS
Single-line charts

- Use single-line charts to illustrate the relationship between two continuous variables, such as how median housing prices change across time (figure 7.8) or how mortality rates vary by year of age. Single-line charts can also be used to show distribution of a continuous variable, like the familiar "bell curve" of the IQ distribution.

As in the other types of xy charts, typically the predictor variable is shown on the x axis and the outcome variable on the y axis, along with their associated units. No legend is needed in a single-line chart

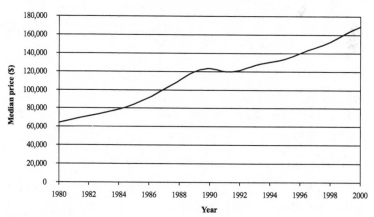

Figure 7.8. Single-line chart
Source: US Census Bureau 2001a

because the two variables are identified in the respective axis labels. If you plot smoothed data (such as moving averages of temporal data, or seasonally adjusted patterns), or other transformations of the original variable, label them accordingly, report those transformations in a footnote, and refer to the text for a more detailed explanation.

Multiple-line charts

Create a multiple-line chart to add a third dimension to a single-line chart. Place the continuous outcome variable on the y axis, the continuous predictor on the x axis, and the categorical predictor in the legend. Use a different line style for each category, with the line styles identified in the legend. For instance, trends in housing prices by region (figure 7.1), with a separate line for each region.

Multiple-line charts with two different y scales

Line charts can also show relations between a continuous variable (on the x axis) and each of two closely related continuous outcomes that are measured in different units (on the y axes). Figure 7.9 shows trends in the number and percentage poor in the United States. Because the units differ for the two outcome variables, one is presented and identified by the title on the left-hand y axis (in this case the percentage of the population that is poor), the other on the right-hand y axis (millions of people in poverty). Use the legend to convey which set of units pertains to each line: the dotted line shows poverty rate and is read from the left-hand y axis (sometimes called the $y1$ axis), while the solid line shows the number of poor persons, on the right-hand ($y2$) axis.

Charts that use two different y axes are complicated to explain and read, so reserve them for relatively savvy audiences and only for variables that are closely related, such as different measures of the same concept. Explicitly refer to the respective y axes' locations and units as you describe the patterns.

xyz line charts

To illustrate the relationship among three continuous variables, such as age, income, and birth weight, create a three-dimensional line chart, sometimes called an xyz chart after the three axes it contains. Label each axis with the name and units of the pertinent variable.

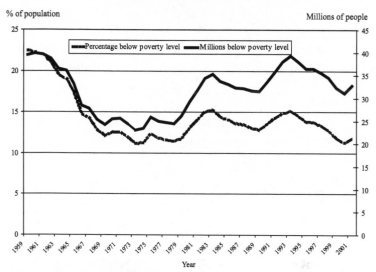

Figure 7.9. Line chart with two y scales
Source: Proctor and Dalaker 2002

HIGH/LOW/CLOSE CHARTS

High/low/close charts present three *y* values for each *x* value. They are probably most familiar as ways to present stock prices, but can also be used to compare distributions, or to present error bars or confidence intervals around point estimates.

Comparing distributions

Use a high/low/close chart to compare the distribution of a continuous variable across categories of an ordinal or nominal variable. In figure 7.10, for example, the median socioeconomic index (SEI) for each of three racial/ethnic groups is plotted by a horizontal dash above its label, with a vertical line showing the first and third quartiles (the interquartile range).

Tukey box-and-whisker plots

Statistician John Tukey developed another variant of this type of chart, now widely known as "box-and-whisker" plots (Tukey 1977; Hoaglin, Mosteller, and Tukey 2000). The "box" portion of the graph shows

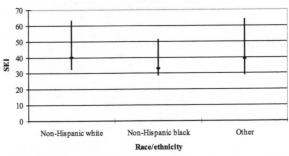

Figure 7.10. High/low/close chart to illustrate median and interquartile range
Source: Davis, Smith, and Marsden 2003

the distance between the first and third quartiles, with the median shown as a dot inside the box. "Whiskers" (vertical lines) extend upward and downward from those points to illustrate the highest and lowest values in the distribution. Outliers—highest or lowest values that are distant from the next closest values—are graphed with asterisks.

Use box-and-whisker techniques to become acquainted with your data, assess how well measures of central tendency (e.g., mean or median) represent the distribution, and identify outliers. Such exploratory graphs are rarely shown in final written documents, but can be invaluable background work to inform how you classify data or choose example values.

Error bars or confidence intervals

Use a high/low/close chart to show error bars or confidence intervals around point estimates;[2] see "Confidence Intervals" in chapter 3 for calculations and interpretation. In figure 7.11, there is one bar color for each racial/ethnic group and one cluster on the *x* axis for each educational attainment group. The height of each bar is the mean birth weight in grams for infants in that group. The 95% confidence interval around each mean birth weight estimate is shown by the vertical error bar that brackets the top of each bar.[3] A chart permits much more rapid comparison of confidence intervals than a tabular presentation of the same numbers because readers can visually examine whether

CIs for different groups overlap one another; see "Explaining a Chart 'Live'" in chapter 12 for an illustrative description.

In formatting high/low/close charts or error bars, ensure that the values to be plotted for each *x* value are measured in consistent units—all in dollars, for example. The *x* variable should be either nominal or ordinal; for a continuous *x* variable, create a line chart with confidence bands. Identify the meaning of the vertical bars either in the title (as in figure 7.11) or a legend.

SCATTER CHARTS

Use scatter charts to depict the relationship between two continuous variables when there is more than one *y* value for each *x* value, such as several observations for each year. A point is plotted for each *x/y* combination in the data, creating a "scatter" rather than a single line. In figure 7.12, each point represents the percentage increase in the adult obesity rate between 1992 and 2002 in one state, plotted against that state's adult obesity rate in 2002. Although both the obesity rate and

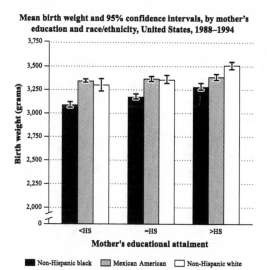

Figure 7.11. Chart to display confidence intervals around point estimates
Source: Miller 2013a, table 15.2.
Note: Data are from the Third US National Health and Nutrition Examination Survey (US DHHS 1997). Weighted to national level using sampling weights provided with the NHANES III (US DHHS 1997).

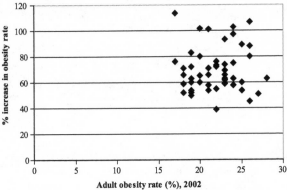

Adult obesity rate (%) versus percentage change in obesity rate, United States, 1992–2002, by state

Figure 7.12. Scatter chart showing association of two continuous variables
Source: Johnson 2004

increase in obesity rate are well above zero in every state, both axis scales start at 0 to avoid misrepresenting the levels of those variables (see "Common Errors in Chart Creation" below).

Scatter charts can be combined with other formats or features, such as to show points from two different sets on the same scatter chart, using different symbols to plot points from each set. For example, an asterisk could be used to show data for men, a pound sign for women.

For a simple scatter chart, no legend is needed because the two variables and their units are identified in the axis titles and labels. For a scatter chart showing more than one series, provide a legend to identify the groups associated with different plotting symbols.

Maps of Numeric Data

Maps are superior to tables or other types of charts for portraying data with a geographic component because they show the spatial arrangement of different cases. They can display most types of quantitative comparison, including level, rank, percentage change, rates, or average values for each geographic unit. For example, figure 7.13 displays average annual pay for each of the lower 48 United States

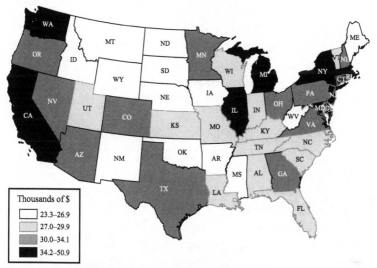

Average annual pay ($1,000s) by state,
United States, 1999

Thousands of $
☐ 23.3–26.9
▨ 27.0–29.9
▨ 30.0–34.1
■ 34.2–50.9

Figure 7.13. Map to illustrate numeric pattern by geographic unit
Source: US Census Bureau 2002c. Average annual pay grouped into quartiles

in 1999, revealing a cluster of states with pay in the top quartile in the Northeast and a cluster of bottom-quartile states in the northern Rockies and upper Midwest. These patterns would be much more difficult to visualize from other types of charts or tabular presentation of data.

Maps can also convey location, distance, and other geographic attributes. Include a legend to explain shading or symbols used on the map, and a ruler to show scale. Most of the basic principles mentioned above for effective titles and layout also apply to maps. See Monmonier (1993) or Slocum (1998) for in-depth guides to using maps to display numeric data.

ADDING DIMENSIONS TO CHARTS

Most of the chart types described above can display relationships among two or at most three variables. Use panels or combine formats to include additional variables.

- To illustrate the age distribution for each of three countries, for instance, create one panel like that in figure 7.3 for each country, then display all the panels on a single page or on facing pages. Use a uniform axis scale and chart size for all panels of the chart to avoid misleading viewers (see "Common Errors in Chart Creation" below).
- To compare changes across time in the distribution of federal outlays for several countries—as in figure 7.7, but with more countries—create a cluster for each year with a stack for each country and slices for each category of outlay.

ADVANCED CHART FEATURES

If your analysis involves comparing your data against standard values or patterns or identifying exceptions to a general pattern, consider using reference points, lines, regions, or annotations in your chart. Include these features only if they are vital to illustrating a pattern or contrast, then refer to them in the accompanying narrative. Many charts work fine without these extra bells and whistles, which can distract your readers and clutter the chart.

Annotations

Annotations include labels to guide readers to specific data values or to provide information to help interpret the chart. On a graph showing a skewed distribution you might report and indicate the median or mean with arrows and a short note or label, for instance. Reserve such annotations for when they convey information that is otherwise not evident from the chart. If the distribution is a symmetric bell curve, keep the graph simple and report median and mean in the text or a table. Annotations can also be used to show outliers or values you use as illustrative examples. In that case, omit data labels from all other values on the chart, naming only those to which you refer in the text.

Another useful type of annotation is denoting statistical significance with symbols accompanied by a footnote to define their meaning. If a chart involves a three-way association, assign different symbols for each contrast, then use a footnote to specify the meaning of each symbol. In figure 7.5, for example, there are two possible contrasts: (a) emergency room use *within income across race* and (b) emergency room use *within race across income*, requiring two different symbols to communicate the patterns of statistical significance. The dagger in figure 7.5a indicates that among the non-poor, odds ratios for blacks and non-blacks are statistically significantly different from one another, while the asterisk in figure 7.5b indicates that among non-blacks, the odds ratio of ER use for non-poor group is statistically significantly different from the odds ratio for the other two income groups. Racial differences in odds ratios for the other two income groups are not statistically significant, nor are differences across income groups among blacks.

Reference Points, Lines, and Regions

Include reference points, lines, or regions in your chart to call attention to one or more important values against which other numbers are to be evaluated.

REFERENCE POINTS

Reference points are probably most familiar from spatial maps. A famous example of the analytic importance of a reference point is the spot map created by John Snow (1936) in his investigation of the London cholera epidemic of 1854. By mapping the location of water pumps where many residents obtained their drinking and cooking water along with the residence of each cholera case, he was able to demonstrate a link between water source and cholera.

REFERENCE LINES

Use reference lines to show the position of a threshold level or other reference value against which to compare data on individual cases or groups.

Horizontal reference lines

On an *xy* chart, a horizontal reference line identifies a value of the outcome variable that has an important substantive interpretation.

- On a scatter chart of blood alcohol levels, cases above a line showing the legal limit would be classified as legally drunk.
- On a bar chart showing ratios (such as odds ratios), a line at

y = 1.0 differentiates cases above and below the reference
value (ratios > 1.0 and < 1.0, respectively).

Vertical reference lines

To identify pertinent values of a continuous independent variable,
include a reference line emanating upward from the x axis. For example, show the date of a change in legislation on a trend chart.

REFERENCE REGIONS

Use a reference region to locate a range of values on the x axis (a vertical region) or y axis (horizontal band) that is relevant to your analysis.
In figure 7.14, shading periods of recession facilitates comparison of
how the number of poor persons increased and decreased during and
between those periods.

Reference regions can also enhance spatial maps: analyses of effects of nuclear accidents like Chernobyl included maps with concen-
tric circles to show geographic extent of different levels of exposure.

Number of poor people, United States, 1959–2001

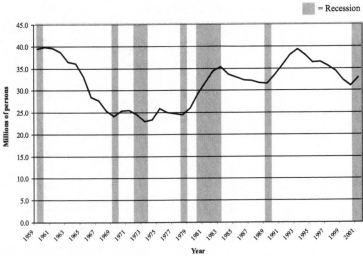

Figure 7.14. Line chart with reference regions
Source: Proctor and Dalaker 2002

In a study of whether to merge local emergency services, you might show which locations can be reached within five minutes by a local police or fire company, which you also plot on the map.

CHOOSING AN APPROPRIATE CHART TYPE

To choose among the many types of charts to present your numbers, first figure out how many and what types of variables you are working with, then consider your audience.

Number and Types of Variables in a Chart

Table 7.1 summarizes the different types of charts for presenting distribution of a single variable, relationships between two variables, and relationships among three variables. Within each of those broad categories, chart types are organized according to whether they involve categorical (nominal or ordinal) or continuous variables, or a mixture of the two types. Start by finding the row that matches the type of task and the number and kinds of variables for your chart (leftmost column), then read across to find suggested chart types with examples of topics and comments on chart design and application.

Unless otherwise noted in the column for example topics, the chart type can accommodate only single-response items; for multiple-response items, a separate chart of that type must be created for each category of response. Accompany charts that present multiple-response items with a footnote explaining that each respondent could have given more than one answer, hence the sum of frequencies across all categories can exceed the number of respondents.

Audience

For nonscientific audiences, keep graphs as simple and familiar as possible. Most people understand pie charts, line charts, and simple bar charts. Reserve complicated three-dimensional charts, logarithmic scales, and charts with two different y scales for audiences that have worked with them before. For speeches, design most slides with simple, focused charts even for scientific audiences; complicated

Table 7.1. Choice of chart type for specific tasks and types of variables

Task	Type of chart	Example topic	Comments
Distribution of one variable			
Nominal with ≤5 categories	Pie	Religious affiliation, major religions	Arrange categories by theoretical criteria or in order of frequency.
Nominal with >5 categories	Simple bar	Religious affiliation, major and minor religious groups	
Ordinal	Histogram	Distribution of letter grades	Arrange categories in numeric order.
Continuous with ≤20 values	Histogram or line	Age distribution in 10-year age-groups	Arrange values in numeric order.
Continuous with >20 values	Line	Distribution of height in inches	Arrange values in numeric order.
Continuous with summary measures of range or variance	Box-and-whisker	Distribution of height in inches	Use to illustrate minimum, maximum, interquartile range, or other summary measures of variance.
Relationships between two variables			
Both categorical	Simple bar	Percentage low birth weight (LBW) by race	For a two-category (dichotomous) outcome variable, show frequency of one value for each categorical predictor; unadjusted or adjusted.[a]
	Clustered bar or clustered histogram	Political party affiliation for young, middle-aged, and elderly adults	For distribution of one variable within each of two or three groups, use a clustered bar to create a histogram for each group.
	Stacked bar	Political party affiliation by 10-year age groups	To compare distribution of a variable across more than three groups of a second variable, use a stacked bar.

One categorical, one continuous	Simple bar	Average math scores by school district	Plot one y value for each x value. Unadjusted or adjusted.[a]
	Bar chart with error bars	Mean birth weight estimates by race/ethnicity and mother's education	Show confidence intervals around means of a continuous outcome variable for each category of a nominal or interval predictor variable.
	High/low/close	Distribution of students' math scores within each of several districts *Multiple response:* Distribution of scores in various subjects within one school district	Illustrate distribution of y values for each x value (e.g., district). *See* Box-and-whisker.
Both continuous	Single line	Trend in monthly unemployment rate	Use a line chart if there is only one y value for each x value or to show a summary (e.g., regression line).
	Scatter	Association between retail price of car and gas mileage for all new 2014 year models	Use a scatter chart to show individual points or more than one y value for each x value.
Relationships among three variables			
All categorical	Two-panel stacked bar	Political party by age group and gender	Use one panel for each value of one categorical predictor variable, one bar for each category of the other predictor variable, and slices for each category of outcome variable.

(continued)

Table 7.1. (continued)

Task	Type of chart	Example topic	Comments
Two categorical, one continuous	Clustered bar	Average math scores by district and type of school	Illustrate a three-way association between two categorical predictor variables and a continuous outcome variable; unadjusted or adjusted.[a]
		Multiple response: average scores for different topics by type of school	Create a separate cluster for each item with bars for each group.
Two continuous, one categorical	Multiple line	Time trend in annual income by age group	Use to illustrate a three-way association between one continuous and one categorical predictor variable and a continuous outcome variable; unadjusted or adjusted.[a]
		Multiple response: trend in income components by age group	To compare multiple-response continuous variables across groups, create a separate line for each response.
	Scatter	Weight by height of low-income children in schools with and without school breakfast	Use different symbols to plot values from the different series.
All continuous	Three-dimensional line (xyz)	Relationships among height, weight, and age of children	Use to illustrate an association among two continuous independent variables and a continuous dependent variable; unadjusted or adjusted.[a]

[a] "Unadjusted" refers to simple bivariate or three-way association; "adjusted" refers to estimates from a multivariate model that controls for other attributes (see Miller 2013a). For adjusted estimates, include a footnote or text description identifying the table from which estimates were taken or listing the other variables controlled in the model.

charts are often hard to read from slides and take longer to explain well.

OTHER CONSIDERATIONS

Design your charts to emphasize the evidence related to your research question rather than drawing undue attention to the structure of the chart itself. Edward Tufte (2001) stresses minimizing "nondata ink" and "ducks"—his terms for excessive labeling and garish features that distract readers from the numeric information itself.

Organizing Charts to Coordinate with Your Writing

Plan charts so the order in which you mention variables or values in the text matches the order in which they appear in the chart, following the principles for coordinating tables with your writing in chapter 6 and Miller (2007a). For ordinal, interval, or ratio variables, retain the natural order of values on your axes. For nominal variables or if you are displaying multiple related items, identify the main point you want to make about the data, then arrange the values on the axes or in the legend accordingly.

To illustrate the importance of thinking about how to organize data to coordinate with your narrative, figure 7.15 displays the same data on consumer expenditures by type of expenditure, arranged in three different ways. Figure 7.15a maintains the order of expenditure categories from the 2002 Consumer Expenditure Survey questionnaire. The bar heights and conceptual content of adjacent categories vary erratically from left to right, so this way of organizing the data actually impedes a well-organized narrative description of the pattern. Readers must zigzag back and forth across the axes to answer important substantive questions such as which categories had the highest expenditures, or how expenditures for necessities compared to those for non-necessities. For similar reasons, alphabetical order of categories (not shown) is a poor organizing principle for coordinating with prose: rarely does one want to write about things in alphabetical order.

To improve synchronization with the associated prose description, figure 7.15b reorganizes the same categories in descending order, making it much easier to follow an empirically structured narrative: which categories had the highest average expenditures? Figure 7.15c first groups items into necessities and non-necessities using the Bureau of Labor Statistics definitions (Duly 2003), and then sorts them in descending order within each of those broad groupings, supporting a coherent discussion of theoretical criteria related to the topic.

Use of Color
Graphics software often automatically uses different colors for each line or bar in a chart. Although slides and some publications permit

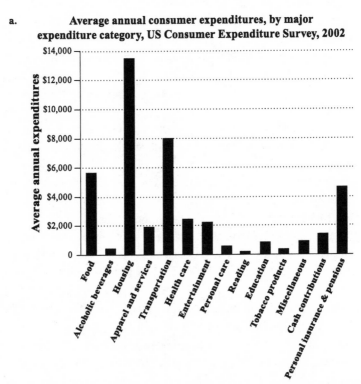

a. **Average annual consumer expenditures, by major expenditure category, US Consumer Expenditure Survey, 2002**

Figure 7.15. Three versions of the same chart to illustrate principles for organizing data: (a) Original order of items, (b) empirical order, (c) theoretical and empirical order
Source: US Department of Labor, Bureau of Labor Statistics 2004.

b.

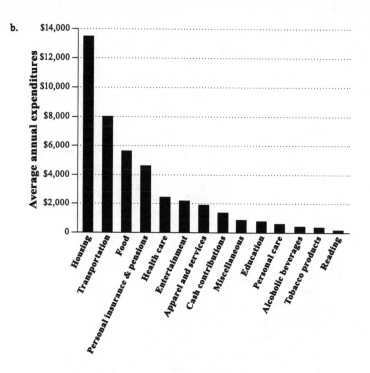

c.

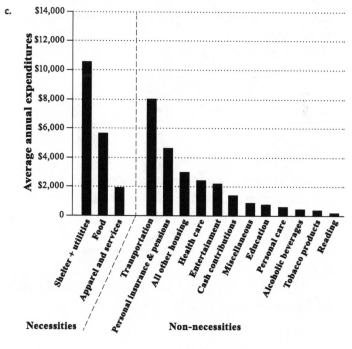

use of color, in many instances you will need to replace colors with different patterns or shades of gray.

- Most documents are printed in grayscale, meaning that the only colors available are black, white, and shades of gray.
- Even if your original will be printed in color, it may be photocopied into black and white.
- Handouts printed from slides are often distributed in black and white.

What appear as different primary hues or pastel tints on a computer screen can become virtually indistinguishable tones in a black-and-white rendition. Choose a color or shading scheme that will remain evident and interpretable regardless of how your chart is reproduced. For color documents or slides, select colors with maximum contrast such as yellow, red, and blue, then make a second version in black and white for other uses. For black-and-white documents, replace colors with one of the following:

- For line charts, pick a different style (solid, dashed, dotted) for each line. If data points are plotted, also vary the plotting symbol, using a circle for one group, a diamond for another, etc.
- For pie charts, bar charts, or maps that include shaded regions, use a different type of shading for each group.
  - If there are only two or three groups (slices or bar colors), use white for one, black for the second, and gray for a third. Avoid using color alone to contrast more than three categories in grayscale, as it is difficult to discriminate among light, medium, and dark gray, particularly if the graph has been photocopied from an original.
  - For four or more slices or bar colors, use different shading schemes (e.g., vertical, horizontal, or diagonal hatching, dots) in combination with solid black, white, and gray to differentiate among the groups to be compared, as in figure 7.2.
- For scatter charts, use a different plotting symbol to plot values for each group. For more than two or three groups create

separate panels of a scatter plot for each group, as patterns for more than three types of symbols become difficult to distinguish on one graph.

Once you have created your chart, print it out to evaluate the color or shading scheme: check that line styles can be differentiated from one another, that different slices or bars don't appear to be the same shade, and that your shading combinations aren't too dizzying.

Three-Dimensional Effects

Many graphing software programs offer the option of making bars or pie slices appear "3-D." Steer clear of these features, which tend to disguise rather than enhance presentation of the data. Your objective is to convey the relative values for different groups or cases in the chart—a task accomplished perfectly well by a flat (two-dimensional) bar or slice. By adding volume to a two-dimensional image, 3-D effects can distort the relative sizes of values by inflating the apparent size of some components. Also avoid tilted or other angled perspectives on pie or bar charts, which can misrepresent proportions. See Tufte (2001) for more details and examples.

Number of Series

For legibility, limit the number of series, particularly if your chart will be printed small. On a multiple-line chart, aim for no more than eight categories (corresponding to eight lines on the chart)—fewer if the lines are close together or cross one another. In a clustered bar chart, consider the total number of bars to be displayed, which equals the number of clusters multiplied by the number of groups in the legend. To avoid overwhelming your readers with too many comparisons, display no more than 20 bars in one chart. An exception: if you will be generalizing a pattern for the entire chart (e.g., "the black bar is higher than the gray bar in every subgroup") with little attention to individual bars, you can get away with more bars.

To display a larger number of comparisons, use theoretical criteria to break them into conceptually related blocks, then make a separate chart or panel for each such block. For example, the questions about knowledge of AIDS transmission shown in figure 7.6 comprise 10 out

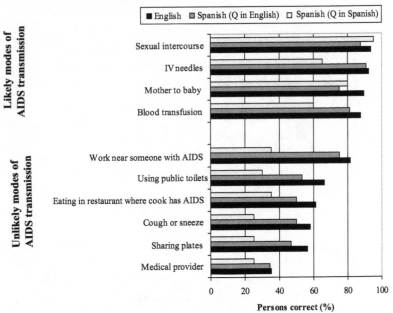

Knowledge of AIDS transmission modes by topic, language spoken at home, and language of questionnaire (Q), New Jersey, 1998

■ English   ▨ Spanish (Q in English)   □ Spanish (Q in Spanish)

Likely modes of AIDS transmission

Sexual intercourse
IV needles
Mother to baby
Blood transfusion

Unlikely modes of AIDS transmission

Work near someone with AIDS
Using public toilets
Eating in restaurant where cook has AIDS
Cough or sneeze
Sharing plates
Medical provider

0   20   40   60   80   100

Persons correct (%)

Figure 7.16. Portrait layout of a clustered bar chart
Source: Miller 2000a

of 17 AIDS knowledge questions from a survey. A separate chart could present patterns of general AIDS knowledge as measured by the other seven questions, which dealt with disease characteristics, symptoms, and treatment. Remember, too, that a chart is best used to convey general impressions, not detailed values; if more precision is needed, substitute a table.

Landscape versus Portrait Layout

Some charts work well with a portrait (vertical) layout rather than the traditional landscape (horizontal) chart layout. Revising figure 7.6 into a portrait layout (figure 7.16) leaves more room to label the AIDS knowledge topics (now on the vertical axis) with clear phrases. Experiment with different pencil-and-paper drafts of your layout before creating the chart on the computer, sketching in the pertinent number of values or groups and the labels for the respective axes.

Linear versus Logarithmic Scale

If the range of values to be plotted spans more than one or two orders of magnitude, consider using a logarithmic scale. On a logarithmic scale, the distance between adjacent tick marks on the axis corresponds to a 10-fold *relative* difference or multiple: in figure 7.17b, for example, the first three tick marks on the *y* axis are for 1, 10, and 100 deaths per 1,000 persons, instead of the uniform 2,000-unit difference between tick marks in figure 7.17a.

Because of the very high mortality rates among the elderly, when mortality rates across the life span are plotted on a linear scale (fig-

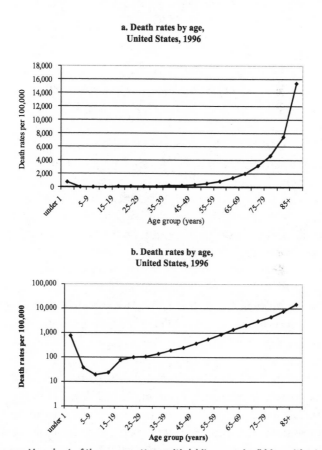

a. Death rates by age,
United States, 1996

b. Death rates by age,
United States, 1996

Figure 7.17. Line chart of the same pattern with (a) linear scale, (b) logarithmic scale
Source: Peters et al. 1998, table 2

ure 7.17a), mortality differences among persons aged 1 to 55 are almost imperceptible, although there is a nearly 40-fold difference between the lowest and highest death rates in that age range. Plotted on a logarithmic scale (figure 7.17b), differences among the low-mortality age groups are easily perceived, yet the much higher mortality rates among the oldest age groups still fit on the graph.

A warning: many nonscientific audiences may not be familiar or comfortable with logarithmic scales, so avoid their use for such readers. If you must use a log scale in those contexts, mention that differences between some values are larger than they appear and use two or three numeric examples to illustrate the range of values before describing the pattern shown in your data.

Digits and Decimal Places

Charts are best used to illustrate general patterns rather than to present exact data values. Choose a level of aggregation with at most five or six digits to avoid illegible axis labels.

COMMON ERRORS IN CHART CREATION

Watch out for some common errors that creep into charts—particularly those produced with computer software. Graphing applications seem to be programmed to visually maximize the difference between displayed values, resulting in misleading axis scales and design that varies across charts. Check your charts for the following design issues before printing your final copy.

Axis Scales

For all variables that include 0 in their plausible range, include 0 on the axis scale to avoid artificially inflating apparent differences. Even a small change can appear huge if the scale begins at a sufficiently high value. In figure 7.18a, the y scale starts at 60%, giving the appearance that the voter participation rate plummeted close to its theoretical minimum, when in fact it was still fully 63% in the latest period shown. When the chart is revised to start the y scale at 0, the possible range of the variable is correctly portrayed and the decline in voter participation appears much more modest (figure 7.18b).

This kind of error crops up most often when presenting small dif-

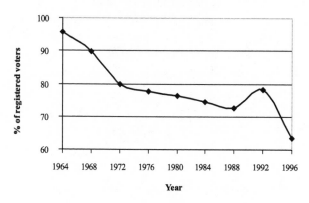

a. Voter participation, US presidential elections, 1964–1996

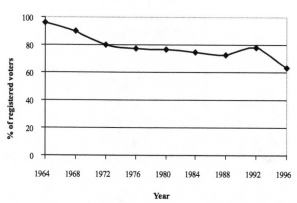

b. Voter participation, US presidential elections, 1964–1996

Figure 7.18. Line charts of same pattern with (a) truncated *y* scale, (b) full *y* scale
Source: International Institute for Democracy and Electoral Assistance 1999

ferences between large numbers. If this is your objective, plot the difference instead of the level, and report the levels elsewhere in the document.

Inconsistent Design of Chart Panels
Use uniform scale and design when creating a series of charts or panels to be compared.

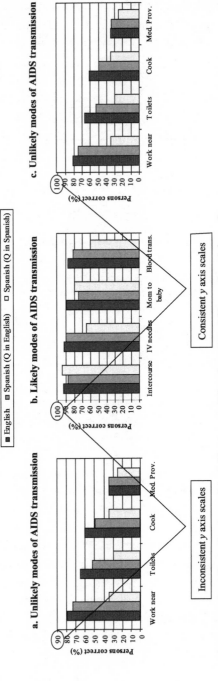

Knowledge of AIDS transmission modes by topic, language spoken at home, and language of questionnaire (Q), New Jersey, 1998

■ English ■ Spanish (Q in English) ▯ Spanish (Q in Spanish)

a. Unlikely modes of AIDS transmission

b. Likely modes of AIDS transmission

c. Unlikely modes of AIDS transmission

Inconsistent y axis scales

Consistent y axis scales

Figure 7.19. Chart illustrating inconsistent (figures a and b) and consistent (figures b and c) y scales
Source: Miller 2000a

Show the same range of y values on each panel: panels a and b of figure 7.19 compare knowledge of unlikely (panel a) and likely (panel b) modes of AIDS transmission. However, the y axis in panel b runs from 0 to 100, while that in panel a runs from 0 to 90. Hence bars that appear the same height in fact represent two very different values. For example, knowledge of "working near someone with AIDS" appears to be as good as knowledge of the four "likely" modes of AIDS transmission, when in fact knowledge of "work near" is lower than all of them. When the scales for the two panels are consistent (panels b and c), the relative knowledge levels are displayed correctly.

Occasionally you will have to use different scales for different panels of a chart to accommodate a much wider range of values in one group than another. If so, point out the difference in scale as you compare the panels in the text.

OTHER DESIGN ELEMENTS TO CHECK

- Consistent sizing of comparable charts. Once you have specified a consistent scale for charts to be compared, ensure that the panels are printed in a uniform size on the page, each occupying a half sheet, for example. If one panel is much smaller, a bar of equivalent numeric value will appear shorter than on a larger chart, misleading readers into thinking that the value displayed is lower than those in other panels of the chart.
- Consistent ordering of nominal variables, both on the chart axes and within the legend. If panel A sorts countries in descending order of the dependent variable but panel B uses regional groupings, the same country might appear in very different places in the two panels. Or, if you intentionally organize the charts differently to address different issues, mention that as you write your description.
- Uniform color or shading scheme (for lines, bars, or slices). Make English-speakers the same color in every panel of a given chart, and if possible, all charts within the same document that compare across language groups.

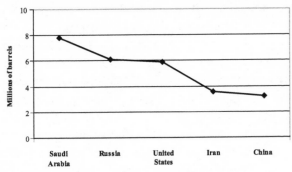

Figure 7.20. Inappropriate use of line chart with nominal data
Source: US National Energy Information Center 2003.

- Consistent line styles (for line charts). If you use a dotted line for the Northeast in panel A, make sure it is represented with a dotted line in panel B.
- Standardized plotting symbols (for scatter charts involving more than one series). If an asterisk represents males in a chart comparing males and females, use that same symbol in other charts that compare the genders.
- Consistent footnote symbols: if an asterisk denotes $p < 0.01$ on figure a, don't use it to indicate $p < 0.05$ on figure b.

Use of Line Charts Where Bar Charts Are Appropriate

LINE CHART FOR NOMINAL VARIABLES

Line charts connect y values for consecutive values on the x axis. Reserve them for interval or ratio data—in other words, for variables that have an inherent numerical order and for which difference can be meaningfully calculated using subtraction. Do not connect values for nominal categories such as crude oil production by country (figure 7.20). The countries were organized in descending order of the dependent variable, but could equally well have followed alphabetical order or grouping by geographic region, so including a connecting line might encourage readers to mistakenly think about a "slope" (e.g., increase in the y variable for a 1-unit increase in the x variable) for this relationship. For nominal variables, use a bar chart (figure 7.4).

LINE CHART FOR UNEQUALLY SPACED ORDINAL CATEGORIES
A more insidious problem is the use of line charts to connect values for ordinal variables that are divided into unequally spaced categories, as with the income data shown in figure 7.21a. Equal spacing of unequal categories misrepresents the true slope of the relationship between the two variables (change in the y variable per 1-unit change in the x variable). To prevent such distortion, treat the x values as interval data by plotting each y value above the midpoint of the corresponding income range on an x axis marked with equal increments (figure 7.21b).

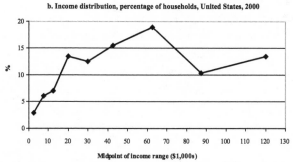

Figure 7.21. Line chart with unequally spaced ordinal variable (a) incorrect x scale, (b) correct x scale
Source: US Census Bureau 2001b

CHECKLIST FOR CREATING EFFECTIVE CHARTS

- To choose an appropriate chart type, compare the attributes of your topic and data to those summarized in table 7.1.

    Consider type of variables.

    Consider number of variables.

    Establish objective of chart, either
    - composition (univariate), or
    - comparison of values of some other variable across groups (bivariate or three-way).

- Evaluate your charts for completeness and consistency.

    Does the title differentiate the topic from those of other charts and tables in the same document?

    Is the chart self-contained? Consider
    - context (W's),
    - units or categories for each variable,
    - legend (if needed),
    - definitions of terms, abbreviations, and symbols
    - data sources.

    Is the chart organized using theoretical and/or empirical criteria, so that it coordinates well with the associated narrative?

    Is the design (color, line style, etc.) consistent with similar charts in the same document?

    Are the axis scale and printed size consistent with other related charts?

    Does the axis scale include zero? If not, does it encompass the plausible range of values for the pertinent variable?

    Is the chart readable in black and white?

CHOOSING EFFECTIVE EXAMPLES
AND ANALOGIES

Examples, analogies, and metaphors are valuable tools for illustrating quantitative findings and concepts. However, choosing effective ones is more complicated than it might first appear. How do you pick analogies that your audience can relate to? How do you avoid selecting numeric contrasts that are too large or too small, or that don't correspond to likely uses of your calculations? As noted in chapter 2, an ideal example is simple, plausible, and relevant to both the issue and audience. Simplicity involves length, familiarity, and wording, while relevance entails considering standard cutoffs and patterns in the field, and other contextual issues. Empirical issues such as the range of values in your data and how the variables are measured should also inform your choice of example. I begin this chapter by describing ways to use examples and analogies in quantitative writing, then present criteria to help you choose effective examples.

WHY USE NUMERIC EXAMPLES?

Every numeric example performs one of several functions: to generate interest in the topic of your work, to quantify differences across groups or time periods, to translate complicated statistical or technical findings into more accessible form, or to illustrate the implications of a statistical finding in a broader social or scientific context.

Establish the Importance of Your Topic
Engage your readers' interest by demonstrating the importance of your topic, ideally right at the beginning of the work. Catch their

attention with a few choice statistics on the frequency with which some problem occurs or the consequences of that phenomenon. If you show that doing something a new way can save them a lot of money or extend their lives by several years, readers will want to find out more about it. If told that scholastic performance is getting worse in their town or that conditions are better elsewhere, they will be motivated to continue reading.

Compare to Previous Statistics

To establish the context and comparability of your findings, relate new information to what is already known. Contrast this year's stock prices with last year's, or compare your findings with results of previous studies. Use quantitative contrasts to assess the extent of similarity or difference between data from different sources. This application of numeric examples is often used in an introduction, a review of the previous literature, or a concluding section of a work.

Illustrate Repercussions of Analytic Results

Use examples to assess substantive significance. Multiply information about proposed property tax changes by the assessed value of a typical home to give residents a sense of how much the new tax will cost them. Combine estimates of reductions in airborne particulates from a new pollution-prevention technology with information on the respiratory effects of particulates to place the new technology in a broader environmental and health perspective. Use these approaches in a general-interest article or in the analytic or concluding section of a scientific paper.

WHY USE ANALOGIES?

Use analogies to help readers grasp the scale of your numbers, understand the shape of an unfamiliar pattern or a relationship between variables, or follow the logic in a multistep calculation by giving them concrete examples of familiar concepts that follow a similar general (abstract) pattern (Willingham 2009). Metaphors, similes, and other related rhetorical devices can also be used to accomplish these tasks; for simplicity I refer to this family of concepts as analogies.

Analogies to Illustrate Scale

Explain very large or very small numbers to audiences that are not conversant with that scale by illustrating those numbers with concrete analogies. To convey the enormity of the world population, the Population Reference Bureau (1999) used the following illustrations of "how much is a billion?"

- "If you were a billion seconds old, you would be 31.7 years of age."
- "The circumference of the earth is 25,000 miles. If you circled the earth 40,000 times, you would have traveled 1 billion miles."

Analogies can also express the scale of very small numbers.

- Translate a very low concentration into the equivalent number of drops of the substance in an Olympic-size swimming pool.
- Provide a benchmark for assessing risk by comparing it to the chances of being struck by lightning.

Other dimensions such as weight, volume, or velocity can also be portrayed with analogies. For a nutrition fact sheet or diet guide, relate standard portion sizes to common objects: a standard 4-ounce serving of meat is equivalent to a regulation-size deck of playing cards; a half-cup serving of rice would fill half a tennis ball.

Analogies to Demonstrate Patterns or Relationships

Portray patterns or relationships using descriptors such as "U-shaped" or "bell-shaped." To illustrate a positive association, compare it to how children's age and height move up together. To describe an inverse association, refer to the relationship between higher prices and lower demand. Analogies can also be used to explain more complicated patterns or relationships. In the business section of the *New York Times*, seasonal adjustment of employment rates was related to the mental process many people apply to the way their body weight changes with the seasons (box 8.1). The fact that the analogy was published shortly after the winter holidays probably only increased its effectiveness.

Analogies to Explain Processes or Calculations

To explicate unfamiliar processes or calculations, relate them to well-known ones. If you liken exponentiating and taking logarithms to "doing and undoing" a mathematical calculation, and follow with a more elementary example of inverse operations such as multiplication and division, most listeners will quickly grasp the basic idea. Descriptions of more complex calculations can also be clarified by comparison to familiar processes, although they often require longer, step-by-step explanations. Consider the following ways to introduce odds ratios to a nonstatistical audience:

> *Poor*: "If the probability of a divorce among people who married as teens is denoted P_t and the probability of a divorce among people who married at older ages is P_o, the odds ratio of divorce for teens compared to older people is the odds for the first group divided by the odds for the second group, or $[P_t/(1-P_t)]/[P_o/(1-P_o)]$."
>
> *An equation full of symbols and subscripts is likely to scare off most nonstatisticians. Even those hardy enough to tackle it will spend a lot of time wading through the notation instead of understanding the logic. For folks who understand odds*

*and odds ratios, just tell them which is the comparison group
(e.g., the denominator); the equation is probably superfluous.*

Not Much Better: "Odds ratios are one set of odds divided by
another. For example, the odds of a divorce are different
depending on age at marriage, so we take the ratio of the odds
in one case (e.g., people who married as teens) to the odds in
the other case (e.g., people who married at older ages). The
ratio of 4/1 to 2/1 equals 2, so the odds ratio is 2.0."

*The basic logic is in place here, but for an audience not used to
thinking about odds, the source of the figures is not clear. Where
did the 4/1 come from? The 2/1? Most people will remember
that a ratio involves division, but wording such as "ratio of ___
to ___" often confuses nonmath folks. Finally, readers are left on
their own to interpret what that value of the odds ratio means.*

Best (for a lay audience): "Odds ratios are a way of comparing
the chances of some event under different circumstances.
Many people are familiar with odds from sports. For example,
if the Yankees beat the Red Sox in two out of three games so
far this season, the odds of another Yankees victory would be
projected as 2-to-1 (two wins against one loss). Now suppose
that the chances of a win depend on who is pitching. Say that
the Yankees won two out of three games against the Red Sox
when Clemens pitched (2-to-1 odds) and two out of four times
when Pettite was on the mound (1-to-1 odds). The *odds ratio*
of a Yankees win for Clemens compared to Pettite is 2-to-1
divided by 1-to-1, or 2. In other words, this measure suggests
that the odds of a Yankees victory are twice as high if Clemens
is the starter. The same logic can be used to estimate how
the relative chances of other types of events differ according
to some characteristic of interest, such as how much odds of
divorce vary by age at marriage."

*This explanation is longer, but every sentence is simple and
explains one step in the logic behind calculating and interpreting
an odds ratio.*

For statistically knowledgeable audiences, you needn't explain the
calculation, but a brief analogy is an effective introductory device:

"An odds ratio measures how the chances of an event change under different conditions, such as the odds of a Yankees victory if Clemens is pitching compared to when Pettite is on the mound."

See "Statistical Significance" in chapter 13 for another example of a metaphor to illustrate a statistical concept.

CRITERIA FOR CHOOSING EXAMPLES AND ANALOGIES

In chapter 2, I introduced two criteria for choosing effective examples: simplicity and plausibility. Here, I elaborate on those criteria and offer several others—familiarity, timeliness, relevance, intended use, and comparability.

Simplicity

Simplicity is in the eye of the beholder, to adapt the old expression. To communicate ideas, choose examples and analogies to fit your audience, not yourself. What seems obvious to one person may be hopelessly obscure to others. A necessarily thorough explanation for one group may be overkill for another. If you are writing for several audiences, adapt the content and wording of your analogies to suit each group. For most lay audiences, the ideal analogy will be as nonquantitative as possible.

FAMILIARITY

Choose analogies that your audience can relate to their own experience. For adults, you might illustrate the penalties associated with missing a deadline by mentioning the consequences of being late for the income tax return filing date. For children, being tardy for school is a better analogy. If your average reader will need to look up a concept or term to grasp your point, find another analogy.

Timeliness increases familiarity. To introduce the field of epidemiology to a group of undergraduates in the mid-1990s, I used the 1976 Legionnaire's disease outbreak at a convention of the Pennsylvania American Legion—the group from which the disease took its name. I subsequently realized most of my students were still in diapers when that outbreak occurred, which is why my example drew a sea of blank stares. A few years later when the movie *Outbreak* was released, students flocked to me to recount scenes from the movie that illustrated various concepts we were learning in class. Now I scan the popular press shortly before I teach to identify fresh examples of the topics I will cover. Especially for a general audience, pick examples that are current or so famous (or infamous) that their salience does not fade with time.

VOCABULARY

Don't obscure your message with a poor choice of words. Use terminology with which your audience is comfortable. Your objective is to communicate, not to demonstrate your own sophistication. Many readers will be completely befuddled by "penultimate" or "sigmoid," but even second graders will understand "second to last" or "S-shaped."

Plausibility

Assessing plausibility requires an intimate acquaintance with both your topic and your data. Don't mindlessly apply values gleaned from other studies, some of which occurred in very different places, times, or groups. For instance, opinions and preferences in a sample of the elderly might not represent those among younger adults or children. Review the literature in your field to learn the theoretical basis of the relationships you are analyzing, then use descriptive statistics and exploratory data analytic techniques to familiarize yourself with the distributions of your key variables and identify unusual values (Miller 2013b).

Sometimes you will want to use both typical and atypical values, to illustrate upper and lower limits or best-case and worst-case scenarios. If so, identify extreme values as such so your audience can differentiate among utopian, draconian, and moderate estimates. See below for more information on sensitivity analyses that compare estimates based on several sets of values or assumptions.

Relevance

A critical facet of a numeric example or comparison is that it be relevant—that it matches its substantive context and likely application.

Before you select numeric values to contrast, identify conventional standards, cutoffs, or comparison values used in the field.

- Evaluations of children's nutritional status often use measures of the number of standard deviations above or below the mean for a standard reference population (Kuczmarski et al. 2000). If you don't use those measures or the same reference population, your findings cannot easily be compared with those of other studies.
- Eligibility for social programs like Medicaid or food stamps is based on multiples of the Federal Poverty Level. If you use purely empirical groupings such as quartiles or standard deviations of the income distribution, your examples will not be directly applicable.

INTENDED USE

Before you choose examples, find out how your intended audience is likely to use the information. Suppose you are studying characteristics that affect responsiveness to a drug rehabilitation program. For an academic audience, you might report estimates from a statistical model that controls simultaneously for age, sex, and educational attainment. For program directors or policy makers, describe patterns for specific age or education groups that correspond to program design features instead.

Comparability

COMPARABILITY OF CONTEXT

In background examples, present data from a similar context (who, what, when, and where). For comparisons, choose data from a context that differs in at most one of those dimensions. If you compare women under age 40 from California in 2010 with people of all ages from the entire United States in 2000, it is hard to know whether any observed differences are due to gender, age, location, or date. Which dimension you vary will depend on the point you want to make: are you examining trends over time, or differences by gender, age group, or location? In each case, cite information that differs only in that dimension, keeping the other W's unchanged.

COMPARABILITY OF UNITS

Make sure the numbers you compare are in consistent units and that those units are familiar to your readers. If you are combining numbers from data sources that report different units, do the conversions before you write, then report all numbers in the same type of unit.

> *Poor*: "The 2012 Toyota Prius hybrid (gas/electric) requires 4.7 liters of gasoline per 100 kilometers, compared to 29 miles per gallon for a 2012 Toyota Corolla with a gasoline-only engine (http://www.fueleconomy.gov/ 2012)."

For some idiosyncratic reason, in the British measurement system fuel economy is reported in terms of distance traveled on a given volume of gasoline (miles per gallon—the higher, the better), but in the metric system the convention is how much gas is required to go a given distance (liters per 100 kilometers—the lower, the better). Hence, this comparison is worse than "apples and oranges." If readers don't pay attention to the units, they will simply compare 4.7 against 29. American and British readers will incorrectly conclude that the gasoline-only engine has better fuel economy, while metric thinkers will conclude that the hybrid is better, but based on incorrect logic.

> *Better* (for those who use British units): "Hybrid (electric/gas) engines improve considerably on fuel economy. For example, the 2012 Toyota Prius hybrid gets 50 miles per gallon (mpg) compared with 29 mpg for the 2012 Toyota Corolla with a gasoline-only engine."

> *Better* (for the rest of the world): "Hybrid (electric/gas) engines improve considerably on fuel economy. For example, the 2012 Toyota Prius hybrid requires only 4.7 liters of gasoline per 100 kilometers (L/100 km) as against 8.1 L/100 km for the 2012 Toyota Corolla with a gasoline-only engine."

Apples to apples, or oranges to oranges, as the case may be.
No need for your readers to conduct a four-step conversion
calculation.

SENSITIVITY ANALYSES

Sensitivity analyses show how results or conclusions vary when different definitions, assumptions, or standards are used. Each definition, assumption, and standard constitutes a different example and is chosen using the criteria outlined above. Sensitivity analyses can be used as follows:

- *To compare results of several different scenarios.* When projecting future population, demographers often generate a series of high, medium, and low projections, each of which assumes a different growth rate. Typically, the medium value is chosen to reflect current conditions, such as the population growth rate from the past year, while the high and low values are plausible higher and lower growth rates.
- *To compare a new standard or definition against its current version.* In the mid-1990s, the National Academy of Sciences convened a panel of experts to assess whether the existing definition of poverty thresholds should be changed to reflect new conditions or knowledge (Citro et al. 1996). Their report included a table showing what poverty rates would have been in each of several demographic groups under both the original and proposed definitions of poverty (table 8.1).

To present detailed results of a sensitivity analysis, create a chart or table like table 8.1. To compare three or more scenarios for a scientific audience, report the different assumptions for each variant in a column of a table (see "Column Spanners" in chapter 6), a footnote, or an appendix. If the definitions, assumptions, or standards are explained in another published source, give a brief description in your document and cite the pertinent source.

Simple comparisons often can be summarized in a sentence or two.

Table 8.1. Tabular presentation of a sensitivity analysis

Poverty rates (%) by group under original and proposed poverty measures, United States, 1992

	Original measure[b]	Proposed measure[a]		Percentage point change Current vs. proposed	
		Alt. 1	Alt. 2	Alt. 1	Alt. 2
Total population	14.5	18.1	19.0	3.6	4.5
Age					
Children <18	21.9	26.4	26.4	4.5	4.5
Adults 65+	12.9	14.6	18.0	1.7	5.1
Race/ethnicity					
White	11.6	15.3	16.1	3.7	4.5
Black	33.2	35.6	36.8	2.4	3.6
Hispanic	29.4	41.0	40.9	10.6	10.5

Source: Citro et al. 1996.

[a] Alternative 1 uses the same income threshold as the original measure, an economy scale factor of 0.75, housing cost index, and a new proposed resource definition. Alternative 2 is the same as Alternative 1 but with an economy scale factor of 0.65. See chapter 5 of Citro et al. 1996 for additional information.

[b] Based on the 1992 threshold of $14,800 for a two-adult/two-child family.

> "In every demographic group examined, estimated poverty rates were several percentage points higher under either Alternative poverty definition 1 or 2 than under the original poverty definition (table 8.1). Changes were larger under Alternative 2 than Alternative 1 in all but one subgroup."

For audiences that aren't interested in the technical details, conduct the analysis behind the scenes and describe the findings in nonstatistical language:

> "Under the current [original] measure, the 1992 poverty rate for the United States as a whole was 14.5%, compared to 18.1% with a new housing cost index and proposed resource definition recommended by an expert panel of the National Academy of Sciences (Citro et al. 1996, 263)."

COMMON PITFALLS IN CHOICE OF NUMERIC EXAMPLES

Failing to examine the distribution of your variables, overlooking measurement issues, falling into one of several decimal system biases, or disregarding default software settings can create problems with your choice of examples.

Ignoring the Distributions of Your Variables

Overlooking the distributions of your variables can lead to some poor choices of numeric examples and contrasts. Armed with information on range, mean, variability, and skewness of the major variables you discuss, you will be in a better position to pick reasonable values, and to be able to characterize them as above or below average, typical or atypical (Miller 2013b). See "Examining the Distribution of Your Variables" in chapter 4.

TYPICAL VALUES

Make sure the examples you intend as illustrations of typical values are in fact typical: avoid using the mean to represent highly skewed distributions or other situations where few cases have the mean value. If Einstein had happened to be one of 10 people randomly chosen to try out a new math aptitude test, the mean score would have vastly overstated the expected performance of an average citizen, so the median or modal value would be a more representative choice. If half the respondents to a public opinion poll strongly agree with a proposed new law and the other half passionately oppose it, characterizing the "average" opinion as in the middle would be inappropriate; in such a case, a key point would be the polarized nature of the distribution.

UNREALISTIC CONTRASTS

Avoid calculating the effect of changing some variable more than it has been observed to vary in real life. Remember, a variable is unlikely to take on the full range of all possible values. Although tape measures include the measurement zero inches, you'd be hard pressed to find anyone that short in a real-world sample. Instead, pick the lowest value found in your data or a low percentile taken from a standard distribution to illustrate the minimum.

If you use the highest and lowest observed values, explain that those values represent upper and lower bounds, then include a couple of smaller contrasts to illustrate more realistic changes. For instance, the reproductive age range for women is biologically fixed at roughly ages 15 to 45 years, on average. However, a woman who is considering childbearing can realistically compare only her current age with older ages, so for women in their 20s, 30s, or 40s, the younger end of that range isn't relevant. Even among teenagers, for whom that entire theoretically possible range is still in their future, few will consider delaying childbearing by 30 years. Hence, comparing the effects of having a child at ages 15 versus 45 isn't likely to apply to many women. A more reasonable contrast would be a five-year difference—age 15 versus 20, or 25 versus 30, for example.

OUT-OF-RANGE VALUES

Take care when using example values that fall outside the range of your data. This issue is probably most familiar for projecting future values based on historical patterns. Accompany your calculations with a description of the assumptions and data upon which they are based.

"The 'medium' population projection assumes an annual rate of growth based on trends in fertility, mortality, and migration during the preceding decade (US Census Bureau 2000)."

Overlooking Measurement Issues

Another issue that affects choice of a numeric contrast is precision of measurement. For example, blood pressure differences of 1 millimeter mercury (mm Hg) are virtually impossible to distinguish with standard sphygmomanometers, and most people cannot report their annual income to the nearest single dollar. If such issues affect any of your variables, choose larger contrasts that can be realistically measured for those variables.

Comparisons also are constrained by how your data were collected. With secondary data you are forced to use someone else's choices of level of aggregation and definitions of categories, whether or not those match your research question. If income data were collected in ranges of $500, you cannot look at effects of smaller changes. Even if you pick values such as $490 and $510 that happen to cross category limits, the real comparison is between the two entire categories (<$500 versus $500–$999), not the $20 difference you appear to be contrasting.

Decimal System Biases

Ours is a decimal (base 10) oriented society: people tend to think in increments of one or multiples of 10, which may or may not correspond to a realistic or interesting contrast for the concepts you are studying. Before you reflexively use a 1- or 10- or 100-unit difference, evaluate whether that difference suits your research question, taking into account theory, the previous literature on the subject, your data, and common usage. Frequently, the "choice" of comparison unit is made by the statistical program used to do the analysis because the default increment often is 1-unit or 10-unit contrasts.

SINGLE-UNIT CONTRASTS

Even if you have concluded that a 1-unit increase is realistic, other contrasts may be of greater interest. Showing how much more food an American family could buy with one dollar more income per week would be a trivial result given today's food prices. A difference of 25 or 50 dollars would be more informative. Even better, find out how much a proposed change in social benefits or the minimum wage would add to weekly income, then examine its effects on food purchases. However, as always, context matters: if you were studying the United States in the early twentieth century or some less developed countries today, that one-dollar contrast in weekly income would be well fitted to the question.

TEN-UNIT CONTRASTS

Some analyses, such as life table calculations, use 10-unit contrasts as the default—a poorly suited choice for many research questions. For instance, infant mortality declines precipitously in the hours, days, and weeks after birth. Ten-day age intervals are too wide to capture mortality variation in the first few weeks of life, and too narrow in the months thereafter. For that topic, more appropriate groupings are the first day of life, the rest of the first week (6 days), the remainder of the first month (21 days), and the rest of the first year (337 days). Although these ranges are of unequal width, mortality is relatively constant within each interval, satisfying an important empirical criterion. Those age ranges also correspond to theory about causes of infant mortality at different ages (Puffer and Serrano 1973; Mathews et al. 2002). Before you choose your contrasts, investigate the appropriate increment for your research question and do the calculations accordingly.

DIGIT PREFERENCE AND HEAPING

Another issue is digit preference, whereby people tend to report certain numeric values more than others nearby because of social convention or out of a preference for particular units. In decimal-oriented societies, folks are apt to prefer numbers that end in 0 or 5 (Barclay 1958), rounding actual values of variables such as weight, age, or income to the nearest multiple of 5 or 10 units. When reporting time, people are inclined to report whole weeks or months rather than exact number of days; or quarter, half, or whole hours rather than exact number of minutes. These patterns result in "heaping"—a higher than expected frequency of those values at the expense of adjacent ones (figure 8.1).

If you have pronounced heaping in your data, treating the heaped responses and those on either side as precise values may be inap-

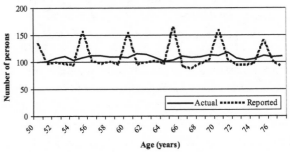

Hypothetical age distribution with and without heaping, ages 50 and older

Figure 8.1. Graphical depiction of data heaping

propriate. Instead, analyze data that are grouped into ranges around those preferred digits. For example, if many people who earn between $23,000 and $27,999 report their income as $25,000, looking at small changes within that range may not make sense. Graphs or tabular frequency distributions can help evaluate whether heaping or digit preference is occurring.

CHECKLIST FOR CHOOSING EFFECTIVE EXAMPLES AND ANALOGIES

- Select analogies or metaphors to fit each intended audience.
    Take into account their knowledge of the topic and concepts.
    Choose ideas and vocabulary that are familiar to them.
- Tailor each numeric example to fit its objective.
    Establish the importance of the topic.
    Compare against previous findings.
    Interpret your statistical results.
    Demonstrate substantive significance of your findings.
- Consider your numeric contrasts.
    Are they within the observed range of values in your data?
    Can that increment be measured reliably given how the data were collected?
    Are your contrasts theoretically plausible for the concepts and units you are studying?
    Are they substantively interesting—neither too big nor too small for real-world conditions?
    Do they apply conventional standards or cutoffs used in the field?
    Do they correspond to likely uses of the results by your intended audience?
- Specify whether the values you present are typical or unusual.
- Evaluate your contrasts.
    Check comparability of context (W's) and units.
    Present one or two selected types of quantitative comparisons.
- For a sensitivity analysis, explain the alternative assumptions or definitions.

Part III

PULLING IT ALL TOGETHER

In the preceding chapters, I described a series of tools and principles for writing about numbers. In practice, rarely will you use these elements piecemeal. Instead, you will integrate them to create a compelling explanation of the issues you address, complete with the quantitative evidence needed to evaluate those questions. Rather than naming or describing the various principles and tools as you use them, you will simply incorporate them into the narrative. The next few chapters show how to do just that, with illustrative "poor/better/best" examples of introductory, data and methods, results, and concluding sections and abstracts for a scientific paper, report, or grant proposal.

In this part, I also demonstrate how to design slides and accompanying speaker's notes for a speech—another common and challenging way to present numeric information. New for this second edition is a chapter on how to design and write materials for nonscientific audiences, including formats such as posters, chartbooks, issue or policy briefs, and general-interest articles. Throughout these chapters, I return often to considerations of audience and format to show how to adapt your writing to suit those varied purposes.

1 2 3 4 5 6 7 8 **9** 10 11 12 13

WRITING ABOUT DISTRIBUTIONS
AND ASSOCIATIONS

Writing about numbers often involves portraying the distribution of a variable, or describing the association between two or more variables. These tasks require several of the principles and tools introduced in the preceding chapters: specifying direction and magnitude of association (chapter 2); considering statistical significance (chapter 3); types of variables, units, and distribution (chapter 4); using quantitative comparisons (chapter 5); organizing the text to coordinate with a table or chart (chapters 6 and 7); and using examples and analogies (chapter 8).

In this chapter, I explain how to combine these elements to write about distributions and associations for scientific audiences, and I describe common types of univariate, bivariate, and three-way patterns. See section on "Presenting Results" in chapter 13 for suggestions on conveying numeric information to applied (nonacademic) audiences.

The strongest descriptions of numeric patterns combine vocabulary or analogies with numeric information because those approaches reinforce one another. Introduce a pattern with a verbal description of its shape and size, then complement it with specific numeric evidence to document that pattern. Well-chosen adjectives, verbs, and adverbs sharpen descriptions of patterns and contrasts, and can add considerable texture and interest. For instance, adjectives like "frigid," "expensive," and "sparse" convey both the topic and approximate numeric value, while verbs such as "plummeted" and "spiked" communicate direction and size of trends (Miller 2010). Phrases and analogies such as "J-shaped" and "bell-shaped" also help evoke general shapes of patterns that can then be fleshed out with results of numeric comparisons such as rank, difference, ratio, or percentage difference to convey level and size of contrasts across values.

WRITING ABOUT DISTRIBUTIONS AND ASSOCIATIONS

Information on distributions of values for single variables, or associations among two or three variables provides the foundation of results sections in scientific articles and is included in many general-interest articles. In an article about elementary education, for instance, you might describe the distribution of class sizes, then show how class size, expenditures per student, and student achievement are related to one another. In a basic statistical analysis or report about experimental data, descriptions of distributions and bivariate or three-way associations often constitute the entire analysis. For more advanced statistical analyses, such as multivariate regressions, those descriptions help demonstrate why a more complex statistical technique is needed (see "Building the Case for a Multivariate Model," in chapter 15 of Miller 2013a).

UNIVARIATE DISTRIBUTIONS

Univariate statistics provide information on one variable at a time, showing how cases are distributed across all possible values of that variable. In a scientific paper, create a table with summary information on each variable, then refer to the table as you describe the data. If you will be comparing across samples or populations, report the frequency distributions using percentages to adjust for differences in the sizes of the samples. One hundred passing scores in a sample of one thousand students reflects a very different share than one hundred passing scores among several million students, for example.

The information you report for a univariate distribution differs for categorical and continuous variables. The type of variable also affects your choice of a chart type to present composition or distribution; see table 7.1 for guidelines.

Categorical Variables

To show composition of a categorical variable, present the frequency of each category as counts (e.g., number of registered Democrats, Republicans, and Independents) or percentages (e.g., the percentage of all registered voters belonging to each party). Report the modal category; the mean (arithmetic average) is meaningless for categorical variables: the "average region" or "mean political affiliation" cannot be calculated or interpreted. If the variable has only a few categories, report numeric information for each category in the text (see examples below).

For variables with more than a handful of categories, create a chart or table to be summarized in the text. Also consider whether two or more small categories might be combined into a larger category without obscuring a group of particular interest to your analysis.

- In some instances, these combinations are based on conceptual similarity. For example, in a comparison of public, private, and parochial schools, you might combine all parochial schools into a single category regardless of religious affiliation.
- In other instances, these combinations are done to avoid tabulating many rare categories separately. For example, you might combine a disparate array of infrequently mentioned ice cream flavors as "other" even though they share little other than their rarity. In these cases, either explain that "other" includes all categories other than those named elsewhere in the table or chart, or include a footnote specifying what the "other" category includes.

The order in which you mention values of a categorical variable depends on your research question and whether that variable is nominal or ordinal. The criteria described on the next few pages also work for organizing descriptions of bivariate or higher-order associations involving categorical variables (see "Associations" below).

NOMINAL VARIABLES
For nominal variables, such as political affiliation or category of federal budget outlays, use principles such as frequency of occurrence or theoretical criteria to organize numbers within a sentence; see "Organizing Tables to Coordinate with Your Writing" in chapter 6. Often it makes sense to mention the most common category and then

describe where the other categories fall, using one or more quantitative comparisons to assess difference in relative shares of different categories.

> *Poor:* "The distribution of US federal budget outlays in 2000 was 61% for human resources, 12% for interest, 16% for national defense, 6% for other functions, and 5% for physical resources (figure 7.2b)."
> *This sentence simply reports the share of federal outlays in each category, requiring readers to do their own comparisons to identify the largest category and to assess the size of differences in the shares of different categories. The categories are inexplicably listed in alphabetical order—a poor choice of organizing principle for most text descriptions.*
> *Better:* "As shown in figure 7.2b, over 60% of US federal budget outlays in 2000 were for human services, with national defense a distant second (16%), and net interest third (12%). The remaining outlays were roughly equally divided between physical resources and other functions."
> *This description uses modal category, rank, and share to convey comparative sizes, reporting numbers only to illustrate relative sizes of the three largest categories.*

If a particular category is of special interest, feature it early in your description even if it isn't the modal value. For example, in a story on the share of interest in the federal budget, highlight that category although it ranks third among the categories of outlays.

ORDINAL VARIABLES

For ordinal variables, the natural sequence of the categories frequently dictates the order in which you report them: young, middle-aged, and elderly age groups, for example. Sometimes, however, you want to emphasize which group is largest or of particular importance in your analysis, suggesting that you mention that group first even if that goes against the usual, ranked sequence of categories. For an article about the influence of the baby boom generation at the turn of the millennium, for instance, discuss that cohort first even though its members were in the middle of the age distribution.

In those situations, begin with a general description of the shape of the distribution, using descriptors such as "bell-shaped" (figures 4.3a and b), "uniform" (figure 4.3c), "U-shaped" (figure 4.3d), or "skewed to the right" (figure 4.3e). Then report frequency of occurrence for the groups you wish to highlight.

Poor: "The age distribution of US adults in 2000 was 8.6% ages 20–24, 8.8% ages 25–29, . . . , and 1.9% ages 85 and older (US Census Bureau 2002d)."
A lengthy list of numbers in the text is overwhelming and a poor way to describe the overall distribution. To present values for each age group, put them in a table or chart.

Poor (version 2): "In 2000, the percentage of US adults who were aged 35–39 was larger than the percentages aged 30–34 or 40–44 (10.3%, 9.3%, and 10.1%, respectively). That age group was also much larger than the oldest age groups (those 80 to 84 and 85+, 2.2% and 1.9%, respectively). Age groups 20–24, 25–29, 45–49 . . . and 75–79 were in between (US Census Bureau 2002d)."
Comparing many pairs of numbers is inefficient and confusing.

Better: "In 2000, the age distribution of US adults was roughly bell-shaped between ages 20 and 55, reflecting the dominant presence of the baby boom cohorts born in the late 1940s through the 1950s (figure 9.1). The largest cohorts were ages 35–39 and 40–44, with 10.3% and 10.1% of the adult population, respectively. After age 55, the age distribution tails off rapidly, revealing the combined effects of smaller birth cohorts and mortality (US Census Bureau 2002d)."
This description uses a familiar shape to summarize the age distribution. By naming the baby boom age groups and mentioning birth cohort size and mortality, it also explains some of the underlying factors that generated the age distribution.

Continuous Variables

For continuous variables such as income, weight, or age, create a table to report the minimum and maximum values, pertinent measure of central tendency, and standard deviation or interquartile range for the variables used in your analysis. If some aspect of the distribution

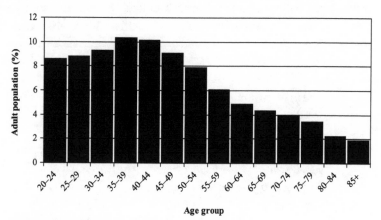

Age distribution of US adults, 2000

Figure 9.1. Age distribution illustrated with a histogram
Source: US Census Bureau 2002d.

affects how you classify data or conduct your analysis, report the frequency distribution and other statistics according to those classifications. For instance, if you compare outcomes for different quartiles of the income distribution, label each category using the pertinent income range (e.g., "lowest quartile: <$15,000") in tables or charts.

In the prose, give summary statistics rather than reporting information on each case or each value, unless there are very few cases. To report fourth-grade class size in a school with four classes at that grade level, for instance, you might list the number of students in each class, followed by the overall average:

> "Kennedy Elementary School opened its doors with fourth-grade classes of 21, 24, 27, and 28 students, for a mean class size of 25 students."

To describe fourth-grade class size for all public schools in a large city, report the range and the mean or median class sizes:

> "On opening day in New York City, fourth-grade class sizes ranged from 20 to 32 students, with an average class size of 25.3."

To provide more detail on the distribution of a continuous variable, create a histogram (to report frequencies for fewer than 20 values) or simple line chart (for 20 or more values). Then portray the general shape of the distribution and report specific values of interest as explained above under "ordinal variables."

Comparing values of a continuous variable against a reference value or cutoff is often informative.

> "The number of elementary school students in classes of 30 or more has tripled in the last three years because of teacher attrition and budget cuts to public schools ... Of current fourth graders [in New York City in 2011–2012], about 14% are in classes of 30 or more students compared with 5.5% during the 2009–2010 school year ... From 2009 [to 2011], the average fourth-grade class grew from 23.4 students to 25.3 students (Phillips 2012)."

ASSOCIATIONS

Most quantitative analyses examine patterns of association between two or more variables. Bivariate patterns describe an association between two variables, such as how mean class size varies by type of school (e.g., public, private, and parochial). Three-way associations introduce a third variable—calculating mean class size by geographic region and type of school simultaneously, for instance. Regardless of the types of variables involved, describe both the direction and magnitude of the association. For scientific audiences, also mention names and results of statistical tests. Morgan et al. (2002) provide an excellent guide to reporting results of many types of statistical tests.

Purpose of Describing Associations

Associations are used to address any of several different objectives.

- They quantify differences in the outcome variable according to values of a predictor variable, such as variation in the percentage of children passing a proficiency test according to school type.
- They describe patterns of association among predictor variables, such as whether different regions have similar distributions of school types.
- They evaluate whether the study sample is representative of the target population, such as whether the demographic

composition of students in the study schools is the same as in all schools.

Roles of Variables and Causal Language

If you are investigating a potential causal relationship, differentiate between the causal variable (the "predictor" or "independent" variable) and the effect (the "outcome" or "dependent" variable) because those roles affect how the statistics are calculated and described. For example, in the relationship between school type and class size, school type is the predictor and class size is the outcome, so report mean class size by school type, not modal school type by class size. For associations among several similar concepts measured at one point in time (e.g., scores in different academic subjects or several concurrent measures of socioeconomic status) or if the causal relationship is ambiguous, avoid causal language; see chapter 3.

Types of Associations

The type of bivariate statistical procedure and associated statistical test depends on whether you are describing an association between two continuous variables, two categorical variables, or one of each. See Moore (1997), Utts (1999), or another statistics text for more background on the underlying statistical concepts.

CORRELATIONS

An association between two continuous variables (e.g., the poverty rate and the unemployment rate) is measured by their correlation coefficient (denoted r). The value of r ranges from -1.0 (for a perfect inverse association) to 1.0 (for a perfect direct association). Variables that are completely uncorrelated have an r of 0.0. Statistical significance is assessed by comparing the correlation coefficient against a critical value, which depends on the number of cases. To describe a correlation, name the two variables and specify the direction of association between them, then report the correlation coefficient and p-value in parentheses:

> "Poverty and unemployment were strongly positively correlated $(r = 0.85; p < 0.01)$."

Bivariate correlations among many variables are usually reported in a table (e.g., table 6.7). Unless you are testing hypotheses about specific pairs of variables, summarize the correlation results rather than writing a detailed narrative about each bivariate association.

> "As expected, the different indicators of academic achievement were highly positively correlated with one another, with Pearson correlation coefficients (r) greater than 0.75 ($p < 0.01$ except between mathematics and language comprehension). Correlations between achievement and aptitude were generally lower."
>
> *This description uses theoretical groupings (academic achievement and aptitude) to simplify the discussion. Generalizations about correlations among measures within and across those classifications replace detailed descriptions of each pairwise correlation, and exceptions are mentioned. One or two specific examples could also be incorporated.*

DIFFERENCES IN MEANS ACROSS GROUPS

An association between a categorical predictor and a continuous outcome variable can be assessed using a difference in means or ANOVA (analysis of variance). F-statistics and t-statistics are used to evaluate statistical significance of an ANOVA and a difference in means, respectively. To describe a relationship between a categorical predictor (e.g., race/ethnicity) and a continuous outcome (e.g., birth weight in grams), report the mean outcome for each category of the predictor in a table or simple bar chart, then explain the pattern in the text.

> *Poor:* "The difference in mean birth weight across racial/ethnic groups was statistically significant ($p < 0.01$; table 9.1).
> *This sentence omits the direction and magnitude of the association.*
>
> *Slightly Better:* "Non-Hispanic black infants weighed on average 246 grams less at birth ($p < 0.01$; table 9.1)."
> *This version reports the direction, size, and statistical significance of the association, but fails to specify the comparison group: less than what (or whom)? As written, this statement could mean*

Table 9.1. Cross-tabulations and differences in means for study variables
Mean birth weight and percentage low birth weight by race/ethnicity, United States, 1988–1994

	Non-Hispanic white (N = 3,733)	Non-Hispanic black (N = 2,968)	Mexican American (N = 3,112)	All racial/ ethnic groups (N = 9,813)
Mean birth weight (grams)[a]	3,426.8	3,181.3	3,357.3	3,379.2
% low birth weight[ab]	5.8	11.3	7.0	6.8

Note: Weighted to population level using weights provided with the NHANES III (US DHHS 1997); sample size is unweighted.
[a] Differences across racial/ethnic origin groups were statistically significant at $p < 0.05$.
[b] Low birth weight: <2,500 grams or 5.5 pounds.

> *that black infants weighed less at birth than at some other age,*
> *that average birth weight for black infants is less now than it was*
> *a few years ago, or that black infants weigh less than some other*
> *group.*
>
> *Best:* "On average, Non-Hispanic black newborns were 246 grams
> lighter than non-Hispanic white newborns (3,181 and
> 3,427 grams, respectively; $p < 0.01$; table 9.1)."
>
> *This description incorporates the mean birth weight, direction*
> *and magnitude, and statistical significance of the birth*
> *weight difference. It also identifies the comparison group and*
> *units of measurement. For a nonacademic audience, omit the*
> *p-value.*

To summarize mean outcomes across several related categorical predictor variables, name the categories with the highest and lowest values of the outcome, then summarize where values for the other categories fall.

> "In the Appleville school district, 2005 SAT II achievement scores
> were highest on the English language test (mean = 530 out
> of 800 possible points) and lowest in mathematics (mean =
> 475). Average scores in science (four subject areas) also fell

below the 500 mark, while those for foreign languages (five languages) and social studies (three subject areas) ranged from 500 to 525."

To avoid mentioning each of the 14 test scores separately, this description combines two criteria to organize the reported numbers: an empirical approach to identify the highest and lowest scores, and substantive criteria to classify the topics into five broad subject areas. Mentioning the highest possible score also helps readers interpret the meaning of the numbers.

Use a consistent set of criteria to organize the table or chart and the associated prose description. To coordinate with the above narrative, create a table that classifies the scores by broad subject area, then use empirical ranking to arrange topics within and across those groupings.

CROSS-TABULATIONS

Cross-tabulations illustrate how the distribution of one categorical variable (e.g., low birth weight status) varies according to categories of a second variable (e.g., race/ethnicity). Statistical significance of differences is assessed by a chi-square test. In addition to showing what percentage of the overall sample was low birth weight, a cross-tabulation reports the percentage low birth weight for infants in each racial/ethnic group. Consider this addition to the above description of birth weight patterns by race:

"These gaps in mean birth weight translate into substantial differences in percentage low birth weight (LBW: <2,500 grams), with nearly twice the risk of LBW among non-Hispanic black as non-Hispanic white infants (11.3% and 5.8%, respectively; p < 0.01, table 9.1)."

The acronym LBW is spelled out and defined at first usage. A ratio is used to quantify the variation in LBW across racial groups, and the percentage LBW is reported for each group to show which numbers were used to compute that ratio. Finally, non-Hispanic white infants are identified as the comparison group.

Two common types of associations among three variables are three-way cross-tabulations and differences in means.

Three-way cross-tabulations

Three-way cross-tabulations are used to investigate patterns among three categorical variables. A cross-tabulation of passing a math test (a yes/no variable), region, and school type would produce information about how many (and what percentage of) students in each combination of region and school type passed the test. As with bivariate cross-tabulations, the test statistic is the chi-square.

Differences in means

Differences in means are used to quantify patterns among one continuous outcome variable (e.g., math scores) and two categorical predictors (e.g., region and school type). If there are four regions and three school types, this procedure would yield 12 average math scores, one for each combination of region and school type. Statistical significance of differences across groups is assessed by an F-test from a two-way ANOVA.

Describing associations among three variables is complicated because three dimensions are involved, generating more values and patterns to report and interpret. To avoid explaining every number or focusing on a few arbitrarily chosen numbers, use the "generalization, example, exceptions" (GEE) technique that was introduced in chapter 2.

"GENERALIZATION, EXAMPLE, EXCEPTIONS" REVISITED

To describe a three-way association, start by identifying the three two-way associations among the variables involved. For example, the relationship between race, gender, and labor force participation encompasses three bivariate associations: (1) race and gender, (2) gender and labor force participation, and (3) race and labor force participation. Only if there are exceptions to the general pattern in one or more of those bivariate associations does the three-way association need to be considered separately.

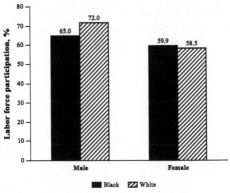

Figure 9.2. Interaction: Exception in direction of association
Source: US Census Bureau 2012.

In figure 9.2, gender and race are predictors of labor force participation (the outcome). If the association between the predictors (e.g., whether the gender distribution differs by race) is important to your study, explain it before discussing how each of the predictors relates to the outcome. Otherwise, focus on the associations between each of the predictors and the outcome. Begin by describing the race/labor force participation relationship and the gender/labor force participation relationship separately, mentioning the three-way pattern only if the two-way associations cannot be generalized:

"(1) In the United States in 2010, labor force participation rates were higher for males than for females, regardless of race (figure 9.2). (2) However, the racial pattern differed by gender, (3) with higher labor force participation among white than black males (72.0% versus 65.0%), but slightly lower labor force participation among white than black females (58.5% versus 59.9%; US Census Bureau 2012) . . ."

The first sentence generalizes the gender pattern in labor force participation, which applies to both races. Phrase 2 explains that the racial pattern cannot be generalized across genders. Phrase 3 describes the direction and magnitude of different parts of the three-way association, documented with numeric evidence from the chart.

In GEE terms, this difference in how race and labor force participation relate is an exception, because no single description of the race/labor force participation pattern fits both genders. In statistical terminology, situations where the association between two variables depends on a third variable are known as *interactions* or *effects modifications* (see Miller 2013a, chapter 16).

Exceptions in Direction, Magnitude, and Statistical Significance
Exceptions can occur in terms of direction, size, or statistical significance of a pattern.

EXCEPTIONS IN DIRECTION
An exception in the *direction* of an association between two groups is fairly easy to detect from a graph of the relationship, as with the pattern of labor force participation by gender and race shown in figure 9.2. Comparing the heights of the respective bars, we observe that male participation is greater than female for both races, but whether black participation is greater than or less than that for whites depends on gender. Exceptions in direction can also occur in trends (e.g., figure 2.1), with a rising trend for one or more groups and a level or declining trend for others.

EXCEPTIONS IN MAGNITUDE
Exceptions in *magnitude* of association are more subtle and difficult to detect. Consider the relationship between gender, race, and life expectancy shown in figure 9.3: the interaction occurs in the different sizes of the gaps between the striped and black bars in the clusters for the two genders.

> "Data from 2010 for the United States show that (1) for both genders, whites outlived blacks (figure 9.3). In addition, (2) females of both races outlive their male counterparts. (3) However, the racial gap is wider for males than for females. Among males, whites outlived blacks by 4.7 years on average, with life expectancies of 76.5 and 71.8 years, respectively. Among females, whites outlived blacks by 3.3 years (life expectancies were 81.3 and 78.0 years; Murphy et al. 2012)."

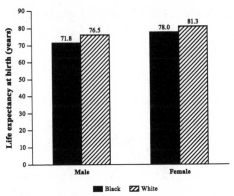

Figure 9.3. Interaction: Exception in magnitude of association
Source: Murphy et al. 2012.

> *In this case, the direction of association in each two-way association can be generalized: the first sentence explains that in all cases, white life expectancy is greater than black, while the second sentence explains that in both racial groups, female life expectancy is greater than male. The sentences in (3) point out the difference in the size of the two "greater thans."*

For trends, exceptions in size appear as a steeper rise or fall for some groups than for others.

EXCEPTIONS IN STATISTICAL SIGNIFICANCE

Generalizations and exceptions also apply to *statistical significance*. If most of the patterns in a table are statistically significant, summarize that finding and note any exceptions. Conversely, if most patterns are *not* statistically significant, generalize, identify the few statistically significant associations as exceptions, and report pertinent test statistics or *p*-values in the text. See box 11.2b for an illustrative example.

Writing about Interactions

The GEE approach is a straightforward way to describe an interaction that is easily understood by most audiences because it emphasizes the substantive patterns before illustrating them with numbers from the associated table or chart. For an audience that is familiar with the statistical meaning of "interaction," use that term as you introduce your results:

> "Race and gender interact in their relation with labor force participation (figure 9.2)."

For guidelines on how to introduce such a pattern to nonscientific audiences, see "Three-Way Associations: Interactions" in chapter 13. Having alerted your audience to the fact that the pattern of association escapes a simple generalization, proceed through the rest of the GEE as in the description of figure 9.2 above. See also appendix A for a systematic approach to identifying and describing patterns for a GEE. As you write a GEE, choose words that differentiate between broad patterns and exceptions.

WORDING FOR GENERALIZATIONS

To introduce a generalization, use expressions like "in general," "typically," or "by and large" to convey that the pattern characterizes many of the numbers you are summarizing. Phrases such as "virtually all," "in the majority of cases," or "roughly three-quarters of" can enhance the summary by conveying approximately what share of individuals, places, or time periods are encompassed by that general pattern. Often you can work the numeric illustration into the same sentence as the generalization by placing the specific numeric value in parentheses after the pertinent phrase: "Virtually all respondents (98%) ..." or "In a slim majority of cases (53%) ... "

WORDING TO INTRODUCE EXCEPTIONS

To introduce exceptions, use phrases such as "an exception [to that pattern]" or "on the other hand," varying the wording to add interest to your descriptions.

> "In seven out of 10 years studied, [pattern and example]....
> However, in the other three years, [contrasting pattern]."

If your exception is literally the opposite of your generalization (e.g., falling rather than rising), consider expressions such as "on the contrary" or "conversely." Then describe the shape of the exception:

"Among males, white labor force participation exceeded black labor force participation. In contrast, among females, black labor force participation was higher than white."

CHECKLIST FOR WRITING ABOUT DISTRIBUTIONS AND ASSOCIATIONS

To describe univariate distribution or composition, consider the type of variable.

- For continuous variables, report minimum, maximum, and mean values and a measure of variation, e.g., standard deviation.
- For categorical variables,

    report modal category and mention selected other categories of interest;

    use one or two types of quantitative comparisons to describe the relative shares of different values;

    coordinate the order in which you mention categories with their sequence in the associated table or chart, using one or more of the organizing criteria described in chapters 6 and 7.

To describe a bivariate association, incorporate the following:

- Direction of association
- Magnitude of association using selected quantitative comparisons (e.g., difference, ratio, or percentage difference)
- Results of statistical tests (e.g., chi-square or t-statistics, and the associated p-values)

To describe a three-way association, use the GEE approach to

- avoid reporting every number or comparison in the text;
- avoid mentioning only arbitrarily chosen numbers;
- describe exceptions of direction, magnitude, and statistical significance.

WRITING ABOUT DATA AND METHODS

An essential part of writing about numbers is a description of the data and methods used to generate your statistics. This information reveals how well your measures match the concepts you wish to study, and how well the analytic methods capture the relationships among your variables—two important issues that affect how your results are interpreted. Suppose you are writing about an evaluation of a new math curriculum. Having explained why students would be expected to perform better under the new curriculum, you report statistics based on a sample that includes some students following the new curriculum and some following the old. Because the data were collected in the real world, the concepts you seek to study might not be captured well by the available variables. Perhaps math performance was measured with a multiple-choice test in only a few classes, and quite a few children were absent on test day, for example.

In addition, statistical methods require certain assumptions that are not always realistic, thus the methods of analysis might not accurately represent the true relationships among your variables. In an observational study with background information on only a few basic demographic attributes, for instance, the assumption of an experiment-like comparison of the two curriculums is unlikely to be satisfied.

To convey the salience of these issues for your work, write about how the study design, measures, and analytic methods suit the research question; how they improve upon previous analyses of the topic; and what questions remain to be answered with other data and different methods. With this information, readers can assess the quality and interpretation of your results, and understand how your analyses contribute to the body of knowledge in the field.

In this chapter I show how to apply the principles and tools covered in previous chapters to writing about data and methods. I begin by discussing how to decide on the appropriate level of detail for your audience and the type of document you are writing. I then give guidelines about the contents of data and methods sections, mentioning many aspects of study design, measurement of variables, and statistical analysis. Finally, I demonstrate how to write about data and methods as you describe the conclusions of your study. In the interest of space, I refer to other sources on these topics. See Chambliss and Schutt (2012) or Lilienfeld and Stolley (1994) for general references on research design, Wilkinson et al. (1999) for a comprehensive guide to data and methods sections for scientific papers, and Miller (2013a) for how to write about data and methods for multivariate analyses.

WHERE AND HOW MUCH TO WRITE
ABOUT DATA AND METHODS

The placement and level of detail about data and methods depend on your audience and the length of your work. For scientific readers, write a dedicated, detailed data and methods section. For readers with an interest in the topic but not the methods, include the basic information as you describe the findings, following the guidelines on "Giving an Overview of Data and Methods" in chapter 13. Regardless of audience, include a discussion of how these issues affect your conclusions. In the paragraphs that follow, I touch briefly on the different objectives of a data and methods section and the discussion of data and methods in a concluding section. Later in the chapter, I give a more detailed look at the respective contents and styles of those sections.

Data and Methods Sections

In articles, books, or grant proposals for scientific audiences, comprehensive, precise information on data sources and statistical methods is expected. In such works, a well-written data and methods section will provide enough information that someone could replicate your analysis: if they were to collect data using your guidelines, they would end up with a comparable study design and variables. If they were to use the same data set and follow your procedures for excluding cases, defining variables, and applying analytic methods, they could reproduce your results.

Data and Methods in the Conclusion

In the concluding section of both lay and scientific papers, emphasize the implications of data and methods for your conclusions. Discuss the strengths and limitations of the data and methods to place your findings in the larger context of what is and isn't known on your topic. Review the potential biases that affect your data, and explain the plusses and minuses of the analytic techniques for your research question and data. For an applied audience, skip the technical details and use everyday language to briefly describe how your data and methods affect interpretation of your key findings.

HOW MUCH TECHNICAL STUFF?

Scientific papers and proposals devote an entire section to data and methods, sometimes called just "methods" or "methods and materials." One of the most difficult aspects of writing these sections—particularly for novices—is selecting the appropriate level of detail. Some beginners are astonished that anyone would care how they conducted their analyses, thinking that only the results matter. Others slavishly report every alternative coding scheme and step of their exploratory analysis, yielding an avalanche of information for readers to sift through. A couple of guiding principles will help you arrive at a happy medium.

First, unless explanation of a particular aspect of the data or methods is needed for your audience to understand your analyses, keep your description brief and refer to other publications that give the pertinent details. If your document is the first to describe a new data collection strategy, measurement approach, or analytic method, thoroughly and systematically report the steps of the new procedure and how they were developed. If the method has been described elsewhere, restrict your explanation to the aspects needed to understand the current analysis, then cite other works for additional information.

Second, conventions about depth and organization of data and methods sections vary by discipline, level of training, and length of the work. To determine the appropriate style for your work, read examples of similar documents for comparable audiences in your field. Some general guidelines for common types of quantitative writing:

- For a journal article or research proposal with a *methodological* emphasis, provide details on methods of data collection or analysis that are novel, including why they are needed and the kinds of data and research questions for which they are best suited.

- For a journal article or research proposal with a *substantive* emphasis, summarize the data and methods concisely and explain the basic logic of your analytic approach, then return to their advantages and disadvantages in the concluding section. Give less prominence to these issues in both the data and methods and discussion sections than you would in a methodological paper.

- For a book or doctoral dissertation, follow the general guidelines above but take advantage of the additional length to provide more detail and documentation. If quantitative methods aren't a major focus, relegate technical details to appendixes or cite other works.

- For a "data book" designed to serve as a reference data source, summarize data and methods in the body of the report, with technical information and pertinent citations in appendixes.

- For documentation to accompany public release of a data set, give a comprehensive, detailed explanation of study design, wording of questions, coding or calculation of variables, imputation, derivation and application of sampling weights, and so forth, accompanied by citations to pertinent methodological documents. Documentation serves as the main reference source for all subsequent users, so dot all the i's and cross all the t's. For excellent examples, see National Center for Health Statistics (1994) and Westat (1994, 1996).

- For general-interest articles or other materials intended for nonstatistical audiences, follow the guidelines in "Giving an Overview of Data and Methods" in chapter 13.

DATA SECTION

More than any other part of a scientific paper, a data and methods section is like a checklist written in sentence form. Organize the description of your data around the W's—who, what, when, where—and two honorary W's, how many and how. If you are using data from a named secondary source such as the Panel Study of Income Dynamics or the National Health Interview Survey, identify that source and provide a citation for published documentation.

When

Specify whether your data pertain to a single time point (cross-sectional data), to different samples compared across several points in time (repeated cross-sections), or to a sample followed over a period of time (longitudinal data), then report the pertinent dates and units of time.

Where

Identify where your data were collected. For studies of human populations, "where" usually encompasses standard geographic units such as cities, countries, or continents, or institutions such as schools, hospitals, or professional organizations. For ecological, geological, or other natural science studies, other types of places (such as bodies of water, landforms, or ecologic zones) or other geographic and topographic attributes (like latitude, longitude, altitude, or depth) may pertain. If geography is important to your topic, include one or more maps to orient an unfamiliar audience.

How

In the data section, "how" encompasses how the data were collected and prepared for analysis, including study design, data sources, wording and calculation of variables, and how missing values on individual variables were handled. ("How" you analyzed the data is the focus of the methods section; see below.) State whether your data are primary data (collected by you or someone on your research team) or secondary data (collected by someone else and not necessarily designed with your research question in mind). For primary data collection involving human subjects, name the institutional review board that evaluated the methodology (for general guidelines, see National Institutes for Health 2002).

STUDY DESIGN

Study designs range from case studies to censuses, from surveys to surveillance systems, from randomized controlled trials to case-control studies, each of which has strengths and weaknesses. In the data section, specify which type of study design was used to collect your data, and then return to its advantages and limitations in the conclusion. See Chambliss and Schutt (2012) or Lilienfeld and Stolley (1994) for more on the topics outlined below.

Indicate whether data were collected cross-sectionally (all variables measured at the same point in time), retrospectively (looking back in time), or prospectively (moving forward in time). State whether the study was a randomized clinical trial or other form of experimental study, a panel study, case-control study, or other type of design. Terminology for study design varies by discipline, so employ the wording used in the field for which you are writing.

Experimental design

If the study involved data from an experiment or randomized controlled trial, explain the experimental conditions:

- Explain how subjects were assigned to the treatment and control groups.
- Describe alternative conditions, whether treatment and control, or different treatments. For each condition, explain what was done, including duration of the experiment, how often observations were made, and other details that would allow readers to replicate the experiment and associated data collection.
- Mention whether placebos were used and whether single- or double-blinding was used.

Sampling

For studies that involve sampling, indicate whether the cases were selected by random sampling, quota sampling, convenience sampling, or some other approach. If random sampling was used, specify whether it was a simple random sample or a more complex design involving stratification, clustering, or disproportionate sampling. For complex sampling designs, explain in your methods section how sampling weights were used (see below) and mention statistical techniques that were used to adjust for the study design (e.g., Westat 1996).

Some types of study design require additional information about identification or selection of cases. Cohort studies sample on the predictor variable, such as selecting smokers and nonsmokers in a prospective cancer study. Case-control studies sample on the outcome, such as selecting people with and without lung cancer and then retrospectively collecting smoking information. If the study involved matching of cases to controls, describe the criteria and methods.

DATA SOURCES

Specify whether your data came from a questionnaire, interview, surveillance system (e.g., vital registration or cancer registry), administrative records (e.g., medical or tax records, voters' registration), observation of subjects, physical examination, or other sources, and cite associated data documentation. List other attributes of data collection that could introduce sample selection bias, coding mistakes, or other types of errors. These issues vary considerably depending on the method of data collection, so read the literature on the methods used to collect your data to anticipate what other information is relevant. A few illustrations:

- If the data are from a survey, was the questionnaire self-administered or from an in-person or telephone interview? Were the data collected orally, online, or in written format?
- If information was extracted from existing records, who identified relevant cases and transcribed data from the forms: a few people specifically trained for the task, or people who normally work with those records but not for research purposes?
- For scientific measurements, the name and characteristics of the measuring instrument (e.g., type of scale, brand of caliper or thermometer) are often required.

Who

"Who" encompasses several dimensions related to how data were collected and whether some cases were omitted from your sample for one or more of the reasons described in the next few sections and under "Item Nonresponse" later in the chapter. Describe the characteristics of the final sample used in your analysis, which may differ from the sample of cases for which data were originally collected. In the sections below, I discuss several key issues that will affect the set of cases you decide to include in your *final analytic sample*, then in a section by that name (below), I explain how to describe the process by which you arrived at that sample, and what it means for interpretation of your findings.

UNIVERSE OR SAMPLE?

Some data sources aim to include the full universe of cases in the place and time specified, others a subset of those cases. In your data section, state whether your data include all cases in the specified place and time as in a census, or a sample of those cases, like a 1% poll of prospective voters. See "Study Design" and "Sampling" sections above for related issues.

STUDY RESPONSE RATE

Few studies succeed in collecting data on all the cases they sought to study. For instance, some subjects who are selected for a study cannot be contacted or refuse to participate; censuses and surveillance systems overlook some individuals. Report the response rate for your data source as a percentage of the intended study sample.

LOSS TO FOLLOW-UP

Studies that follow subjects across time typically lose some cases between the beginning and end of the study. For instance, studies tracking students' scholastic performance across time lose students who transfer schools, drop out, or refuse to participate. Perhaps you started following a cohort of 500 entering ninth graders, but only 250 remain four years later.

Loss to follow-up, also known as attrition, affects statistical analyses in two important ways. First, the smaller number of cases can affect the power of your statistical tests (Kraemer 1987; Murphy 2010).

Second, if those who are lost differ from those who remain, inferences drawn from analysis of the remaining cases may not be generalizable to the intended universe. For example, those who stay in school are often stronger students than those who drop out, yielding a biased look at the performance of the overall cohort.

For longitudinal data, provide the following information, and then discuss representativeness:

- The number of cases at baseline (the start date of the study), following the guidelines under "Final Analytic Sample" below.
- The number and percentage of the baseline sample that remained at the latest date from which you use data.
- Reasons for dropping out of the study (e.g., moved away, dropped out of school, died), if known, and the number of cases for each.
- The percentage of all baseline cases who are present at each subsequent round, if you use data from intervening time points, e.g., five sets of observations collected at annual intervals.

A chart can be a very effective way to present this information. For example, in a study of religious practices and beliefs in respondents' last year of life, Idler and colleagues (2001) graphed the total number of respondents remaining in each of the four survey years (the downward-sloping line in figure 10.1) and the number of deaths that occurred in a year following a face-to-face interview (in the brackets). The chart complements the accompanying description of study design and sample size by illustrating both the criteria for identifying cases for their study and the overall pattern of attrition.[1]

RESTRICTIONS ON THE ANALYTIC SAMPLE
TO FIT YOUR RESEARCH QUESTION
Once you have reported the overall response rate for the source from which you draw your data, consider whether you must limit the cases used for your analysis to a subset of the overall sample from the original study in order to suit your research question. There are several reasons why this might be necessary, especially if you are using secondary data: inclusion criteria, study design, and sample size issues.

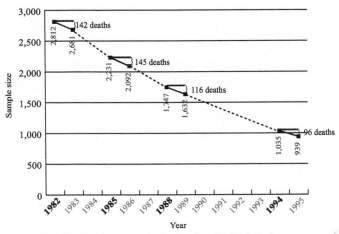

Overall sample size and number of deaths occurring in the year following in-person interviews, New Haven EPESE sample, 1982–1995

Figure 10.1. Line chart to show sample size in a longitudinal study
Source: Idler et al. 2001.
Note: Data are from New Haven EPESE (Cornoni-Huntley et al. 1986). Bracketed portions of survival curve represent deaths in the 12 months immediately following face-to-face interviews, *N* = 499. Bolded dates indicate survey years.

Cases that meet specific criteria for study variables

For some topics, it might be appropriate to restrict your analytic sample to cases that meet certain criteria, such as having particular demographic traits, minimum test scores, or a specific disease. If you excluded cases from your analytic sample for any of these types of reasons, describe those restrictions, explain why they suit your research question, and report the number of cases that remain in the sample after this exclusion.

Missing by design

In some data sources, one or more questions might be skipped by design for some cases in the sample. This might be done for any of the following reasons:

- One or more items might not apply to certain individuals. In these situations, individuals' responses to a *filter question* are used to determine which of them are asked the set of *contingent questions* that pertain only a subset of respondents (Chambliss and Schutt 2012); for others, the contingent questions are not applicable. For instance, any respondents who answer that they were employed throughout the past year would not be asked questions related to dates, duration, or reasons for unemployment. The resulting *skip pattern* means that responses to the contingent questions are only available for the subgroup of respondents to whom they pertain; those variables will be coded "not applicable" for others.

- Some topics might not be asked of specific subgroups of respondents due to concern about their ability to answer the question accurately. For instance, in the NHANES III, birth weight data were collected only for children under age 10 years in order to minimize recall bias in their mother's response (see box 10.1). As in the previous example, a skip pattern was created so that valid responses to that question were available only for a subgroup of all cases in the sample.

- Surveys sometimes administer specialized topic modules only to a randomly selected subsample of respondents, often as a way to obtain a smaller representative sample that meets requirements for statistical power while reducing the cost of the overall study. A core set of questions is asked of all respondents, and then each of several alternative modules is administered to different subsets of the overall sample. For instance, the National Immunization Survey (NIS) used this *split sample mini-module* approach to obtain suitably precise estimates for each of three topic modules: "Ability to Pay for Vaccinations" (asked of 74% of the overall sample), "Parental Knowledge and Experiences" (asked of a different 13% of the sample), and "Day Care/Breastfeeding/WIC" (13%) (Smith et al. 2000). If you were studying breastfeeding using these data, you would be limited to the 13% of the overall sample who were asked questions on that topic.

Read the data documentation, questionnaires, and codebook for your data source to learn whether any of these issues pertains to one or more of the variables used in your analysis. If so, describe the sampling design or skip pattern, report how it affects the number and characteristics of cases for whom those variables are available for your analysis, and provide a citation to the data documentation. A carefully-coded data set will have distinct codes for each variable to differentiate between "not applicable," missing due to item nonresponse (see section below), and valid values.

Minimum subgroup sample size

In some data sets, there might not be enough cases in one or more subgroups of a key variable to provide sufficient statistical power (Kraemer 1987), and it might not be suitable to combine them with other subgroups used in your analysis. For instance, the birth weight analysis described in box 10.1 (below) omitted children from "other" racial/ethnic groups because there were not enough of them to analyze as a separate group. Moreover, theoretical reasons suggested that they should not be combined with any of the other racial/ethnic groups being analyzed (in this case, non-Hispanic white, non-Hispanic black, and Mexican American). If you excluded certain subgroups from your analysis based on sample size considerations, explain the basis for your decision and report the sample size after that exclusion.

What

Having reported the context, study design, and data sources, describe what variables were measured. If all your variables come from the same source, summarize that information in one sentence, and don't repeat. If more than one data source is involved, generalize as much as possible about which variables came from which sources. Use panels within tables or create separate tables to organize variables by topic and source, with information about sources of each variable in the title or footnotes.

Poor: "Age, sex, race, marital status, number of children, income, and educational attainment were taken from the demographic section of the questionnaire. Attitudes about [long list] were taken from the attitudinal section of the questionnaire. Medical records provided information about [long list of health items]. Asthma was also asked about on the questionnaire."

This description is unnecessarily long, repeating information that is far more easily organized in a table.

Better: "Demographic characteristics and attitudinal factors (table A) were drawn from the questionnaire, and most health indicators from the medical records (table B). An exception was asthma, for which information was available from both sources."

This description coordinates with tables (not shown) that organize variables by data source and major conceptual groupings, eliminating the need to specify the source in the text for each variable individually.

VARIABLES

Except for the most detailed scientific texts or data documentation, limit in-depth descriptions of measures to your outcome and key predictor variables. For other, less central variables, mention them as you describe the chart or table in which they are first shown, or refer to another publication that contains information about their attributes. The documentation for the NHANES III—the data used in the birth weight analysis reported here—includes a CD-ROM with copies of the field manuals used by data collectors, spelling out in great detail how various aspects of the health examination, nutritional history, and cognitive testing were conducted (NCHS 1994; Westat 1996). Because that information is publicly available, scientific papers based on those data can refer to that source for details.

Raw versus calculated data

Some of your variables may be used exactly as they were collected (raw data), others calculated from one or more variables in the raw data. For primary data, indicate whether you adopted or adapted the items or scales from other sources or developed your own. For either primary or secondary data, explain whether and how items were pretested, and report on reliability and validity (see below).

For variables analyzed in the same units or categories in which they were originally collected, mention their source.

> *Poor:* "One asthma measure was collected on the mother's questionnaire, the other from medical records."
> *The questionnaire and medical records could have collected data in any of several ways, each of which has different potential biases, so for most scientific papers a more precise description is needed.*
>
> *Better:* "A maternal report of asthma was based on the question 'Have you ever been told your child has asthma?' A doctor's report of asthma was based on (1) checking that diagnosis on a list of possible diagnoses, (2) listing 'asthma' on the open-ended section of the medical record, or (3) listing an IDC9 code of 493 on the open-ended section of the medical record (NCHS 1995)."

If the phrasing of a question could affect how subjects interpreted that question, include the original wording either in a short paragraph within the data section or in an appendix that displays the relevant portion of the data collection instrument; see table 10.1 for an example. The specific wording of items can affect subjects' responses, so avoid rephrasing the original, such as substituting "better than average" for "excellent" and "very good," or replacing "HIV" with "AIDS." Wording of very short items can sometimes be incorporated into table row or column headings (see table 6.2).

To address some research questions, new variables must be created from variables in the original data (Miller 2014). Examples include,

- calculating a categorical variable from a continuous one (e.g., an indicator of whether birth weight fell below the cutoff to define "low birth weight," or quartiles of the income distribution),
- combining categories (e.g., collapsing five-year age groups into 10-year groups),
- creating a summary variable to aggregate information on several related items (e.g., calculating total family income

from wages, salary, alimony, Supplemental Security Insurance benefits, etc.),

- creating a scale (e.g., the CESD scale from a series of items related to depressive symptoms).

Explain how new variables were calculated, whether by you or by those who prepared the secondary data, and mention whether that approach is consistent with other analyses. State whether the criteria used in those calculations were based on existing standards or cut-offs (such as the definition of low birth weight, or the list of income components used to calculate overall income), empirical analysis of your data (such as quartiles of the income distribution), or theoretical criteria, then cite the pertinent literature.

If you will be analyzing an index or scale, list the set of items that comprise it, how they were originally coded, whether reverse coding was used for any items, whether they were summed or averaged to create the index, and any weighting of items (Treiman 2009). If the list of items or questionnaire instructions is long, they can be presented in an appendix such as table 10.1.

Units and categories

Name the subgroups of each categorical variable and the units of measurement for every continuous variable in each table or chart where that variable is reported. Explain units of measurement in the text of the data section only if they are unusual or complex. Mention variables measured in familiar units as you write up corresponding results. Refer to categories of nominal or ordinal variables by their names rather than by the numeric codes (values) assigned to those categories in your database. Accompany ordinal values that were created from continuous variables with their numeric equivalents, e.g.,

"'High' and 'low' correspond to the top and bottom quintiles of the income distribution, while 'middle' comprises the middle three quintiles."

Some ordinal variables will have both units and categories, e.g., age group in years, or income range in Euros. If ordinal variables were

Table 10.1. Grid to show wording and coding of items used to construct a scale

Items about abortion attitudes, 2000 US General Social Survey

"Please tell me whether or not you think it should be possible for a pregnant woman to obtain a legal abortion ... " Circle one code for each statement.	Yes	No	Don't know
If there is a strong chance of serious defect in the baby?	1	2	8
If she is married and does not want any more children?	1	2	8
If the woman's own health is seriously endangered by the pregnancy?	1	2	8
If she is not married and does not want to marry the man?	1	2	8
If she became pregnant as a result of rape?	1	2	8
If she is not married and does not want to marry the man?	1	2	8
If the woman wants it for any reason?	1	2	8

Source: National Opinion Research Center 2000.
Note: Values of "8" becomes missing values in database.

collected without reference to specific numeric values, list the possible responses exactly as they were worded on the original data collection instrument, e.g.,

"How would you rate your health: excellent, very good, good, fair, or poor?"

Outliers

Occasionally, your data will contain "outliers"—values that fall well outside the range of the other values and can substantially affect estimates based on the full sample. To avoid biasing results, outliers are sometimes excluded from an analysis sample. If an NFL first-round draft pick happened to be part of your sample of 100 recent college graduates, you would be well justified excluding him from an analysis of income to avoid grossly inflating average income for typical recent graduates. (If your focus were on NFL salaries, you would choose your sample differently, but that's another story.)

Any omission of selected cases is an opportunity to "finesse" your data to obtain some desired results—in other words, to lie with statistics—so exercise great care in deciding which cases to exclude and in communicating what you did and why. State how outliers were identified and what criteria you used to exclude cases, then report how many cases were excluded. Finally, report how much the size and statistical significance of associations differ with and without the outliers.

Reliability and validity

Indicators of reliability are used to evaluate consistency of alternate measures of the same concept—whether different questions, different observers, or different time points. Measures of validity consider whether a question or scale captures the concepts it is intended to measure, including face validity, concurrent validity, predictive validity, and construct validity. Report standard statistics on the reliability and validity of your key variables. See Chambliss and Schutt (2012) or Treiman (2009) for an overview of reliability and validity; Morgan et al. (2002) for illustrative wording to report the measures.

ITEM NONRESPONSE

Among those who responded to a survey, one or more items might be missing for that respondent, for instance, when a respondent returned the survey but refused to report income, or when information on height was missing from some medical records. This phenomenon is known as "item nonresponse" to differentiate it from study nonresponse. See section on "Restrictions on the Analytic Sample to Fit Your Research Question" above for reasons why some variables might be missing for entire subgroups of respondents.

Cases that are missing the outcome variable or a key predictor variable cannot be used to analyze the association between those variables. Report frequency of item nonresponse that affects more than a few cases and explain how you dealt with missing values on individual items (see Westat 1994 for discussion), especially those that are key measures for your research question. Also report the item response rate, clearly specifying "percentage of what" (chapter 5): for item nonresponse, it should be the percentage of eligible respondents (those who were asked the question) who provided a valid response.

Values missing due to item nonresponse can be handled by omitting those cases from the overall analysis, defining a missing value category, or imputing values.

Missing value category

One approach to handling missing values for a variable is to create a category called "missing." This method is best used to avoid excluding cases that lack data on one of several background variables, and only if a small share of cases have missing values on any one variable. Report the percentage of cases in the missing value category of each variable in a table of descriptive statistics. Comment on its interpretation in the discussion section if it affects more than a small percentage of cases.

Imputation

Imputation involves filling in values for cases that lack data on a variable based on values of that variable or related variables for cases with valid data. See Westat (1994) for an overview of imputation processes and evaluation.

Final Analytic Sample

Once you have considered whether any of the above issues affect the number of cases for whom values of your key variables are available, describe the exclusion criteria used to arrive at your final analytic sample, report the final sample size, and discuss whether your final analytic sample is representative of the population to whom you want to generalize the results (see section on "Data and Methods in the Discussion Section" below).

EXCLUSION CRITERIA

Describe the criteria you used to exclude cases from your analytic sample, such as those due to study design issues, item nonresponse, outlier values, or criteria for minimum subsample size, and explain how they pertain to your particular research question. State how many and what percentage of cases were dropped as a result of each restriction, especially reasons that relate to your key predictor and outcome variables.

SAMPLE SIZE

Report the number of cases in your final analytic sample and the response rate as a percentage of cases to whom the questions pertained, overall and for any major subgroups being compared. If there are only a handful of sample sizes to report (e.g., the total and for each of four age/sex groups), mention them in the text and in tables that report descriptive statistics or analyses. For larger numbers of subsamples (e.g., each of the 50 states), report the smallest, largest, and average subsample sizes, or report subsample sizes in an appendix table. If you are using sampling weights in your analyses (see below), the sample size is the unweighted number of cases.

REPRESENTATIVENESS

Describe how well your sample reflects the population it is intended to represent. Limit this comparison to those who qualify in terms of place, time, and other characteristics that pertain to your research question. For instance, if studying factors that influence urban school performance, don't count students from suburban or rural schools among the excluded cases when assessing representativeness—they aren't part of the population to whom you want to generalize your results.

Depending on your audience and the length of your document, there are several ways to report on representativeness.

- At a minimum, report how many and what percentage of sampled cases were included. If the response rate is over 85%, no additional discussion of representativeness may be needed.
- Create a table of bivariate statistics comparing known characteristics of included and excluded cases, or, if statistics are available from the census or other sources, comparing the sample (weighted to the population level; see table 6.5) to

BOX 10.1. DATA SECTION FOR A SCIENTIFIC PAPER

"(1) Data were extracted from the 1988–1994 National Health and Nutrition Examination Survey III (NHANES III)—a cross-sectional, population-based sample survey of the noninstitutionalized population of the United States (US DHHS 1997). (2) To reduce recall bias, birth weight questions were asked only about children aged 0 to 10 years at the time of the survey. (3) The final analytic sample comprised 3,733 non-Hispanic white, 2,968 non-Hispanic black, and 3,112 Mexican American infants for whom family income and race/ethnicity were known (93% of that age group in the NHANES III). (4) Children of other racial/ethnic backgrounds (mostly Asian) were excluded because there were not enough cases to study them as a separate group.

"(5) All variables used in this analysis were based on information from the interview of the reporting adult in the household. (6) Consistent with World Health Organization conventions (1950) a child was considered to be low birth weight (LBW) if their reported birth weight fell below 2,500 grams or 5.5 lbs. (7) Coding and units of variables are shown in table 6.5, (8) which compares the sample to all US births."

COMMENTS

(1) Names the data source, including dates, specifies type of study design and target population (where, who), and provides a citation to study documentation.

(2) Mentions potential bias for one of the variables and how it was minimized by the study design (age restrictions).

(3) Reports the unweighted sample size for the three major subgroups in the study, specifies exclusion criteria, and reports the associated response rate.

(4) Explains why children of "other" racial/ethnic background were excluded from this birth weight analysis. (For other research questions, different criteria would pertain.)

(5) Specifies which sources from the NHANES III provided data for the variables in the analysis. (The NHANES III also included a

medical history, physical examination, lab tests, and dietary intake history. Those sources were not used in the birth weight analysis described here.)

(6) Explains how the low birth weight indicator was calculated and gives its acronym, with reference to the standard definition and a pertinent reference.

(7) Directs readers to the table of descriptive statistics for details on units and coding of variables, averting lengthy discussion in the text.

(8) Refers to the table with data to assess the representativeness of the sample.

the target population. Statistically significant differences in these traits help identify likely direction of bias, which you can summarize in the concluding section of the paper.

- Describe reasons for exclusion or attrition, such as whether those who remained in the sample differed from who were lost from the sample by moving away, dropping out of school, or dying.

Box 10.1 presents an example of a data section for a scientific paper, with phrases numbered to coordinate with the accompanying comments.

METHODS SECTION

In the methods section of a scientific paper, explain how you analyzed your data, including which statistical methods were used and why, and whether you used sampling weights.

Types of Statistical Methods

Name the statistical methods used to analyze your data (e.g., analysis of variance, Pearson correlation, chi-square test) and the type of software (e.g., SAS, SPSS, Stata) used to estimate those statistics.

Sampling Weights

If your data are from a random sample, they are usually accompanied by sampling weights that reflect the sampling rate, clustering, or disproportionate sampling, and which correct for differential nonresponse and loss to follow-up (Westat 1996, sec. 2). Most analyses of such data use the weights to inflate the estimates back to the population level, estimating the total number of low birth weight infants in the United States based on the number in the study sample, for instance. If disproportionate sampling was used, the data must be weighted to correctly represent the composition of the population from which the sample was drawn (e.g., Westat 1996, sec. 1.3). The sample size used to calculate the standard errors for statistical tests should be the unweighted number of cases. Verify that your statistical program makes this correction.

Explain when and how you used the sampling weights and refer to the data documentation for background information about the derivation and application of those weights. For analyses of clustered data, name the statistical methods used to correct the standard errors for clustering (e.g., Shah et al. 1996).

Equations

Equations are an expected feature in economics, statistics, and related fields as a succinct and efficient way to convey calculations involving several variables or a series of mathematical operations. Using standard notation such as subscripts, superscripts, and other mathematical notation, an equation is often better than sentences for conveying how a measure is calculated. For instance, the equation for computing body mass index (BMI) from height and weight is:

$$BMI = \text{body mass in kilograms}/(\text{height in meters})^2$$

Other examples of the use of equations can be found under "Confidence Intervals" (chapter 3) and "Attributable Risk" (chapter 5).

For many situations and audiences, however, unless the statistical method is new (or at least new to your field or topic) or involves a complex calculation, equations are often superfluous. Before including equations for nonscientific audiences, consider whether they con-

"(1) Bivariate associations among race/ethnicity, socioeconomic characteristics, and birth weight were tested using *t*-tests [for continuous outcomes such as birth weight in grams] and chi-square tests [for categorical outcomes such as low birth weight]. (2) All statistics were weighted to the national level using weights provided for the NHANES III by the National Center for Health Statistics (US DHHS 1997). (3) SUDAAN software was used to adjust the estimated standard errors for complex sampling design (Shah et al. 1996)."

COMMENTS

(1) Identifies types of statistics used to test bivariate associations. Text shown in brackets clarifies which statistical tests pertain to which outcomes, and would be included only for students of elementary statistics.

(2) Mentions use and source of sampling weights and cites pertinent documentation.

(3) Explains what method and software were used to correct for complex study design, with a citation.

tribute important information not otherwise available in your document or other publications. For lay readers or academic audiences of nonstatisticians, explain the statistical *concepts* embodied by the method, paraphrasing jargon into less technical terminology of the field into which the research question fits, and using equations as sparingly as possible. See "Giving an Overview of Data and Methods" in chapter 13 for examples.

Box 10.2 is an illustrative methods section for a scientific paper about the birth weight analysis, with phrases numbered to coordinate with the associated comments.

DATA AND METHODS IN THE DISCUSSION SECTION

Many of the elements described in the data and methods sections have repercussions for the analysis and interpretation of your results.

Explain these issues by discussing the advantages and limitations of your data and methods in your conclusion.

Study Strengths

Remind your audience how your analysis contributes to the existing literature with a discussion of the strengths of your data and methods. Point out if you used more recent information than previous studies, data on the particular geographic or demographic group of interest, or improved measures of key variables, for example. Mention any methodological advances or analytic techniques you used that better suit the research question or data, such as taking a potential confounding variable into account.

Rather than repeat the W's and other technical details from the data and methods section, rephrase them to emphasize specifically how those attributes strengthen your analyses.

> Poor: "The experimental nature of the study strengthened the
> findings by eliminating self-selection."
> *This generality about experimental studies doesn't convey how*
> *they affected this particular research question and data source.*
> Better: "Because subjects were randomly assigned to the
> treatment and control groups, differences in background
> characteristics of those who received treatment were ruled out
> as an explanation for better survival in that group."
> *This version explains how an experimental design*
> *(randomization) improved this study, mentioning the outcome*
> *(survival), the predictor (treatment versus control) and potential*
> *confounders (background characteristics).*

Study Limitations

Just as important as touting the strengths of your data and methods is confessing their limitations. Many neophytes flinch at this suggestion, fearing that they will undermine their own work if they identify its weaknesses. On the contrary, part of demonstrating that you have done good science is acknowledging that no study is perfect for answering all questions. Having already pointed out the strengths of your study, list aspects of your analyses that should be reexamined with better data or statistical methods. Translate general statements about biases or other limitations into specific points about how they affect interpretation of your findings.

> *Poor:* "The study sample is not representative, hence results cannot be generalized to the overall target population." *This statement is so broad that it doesn't convey direction of possible biases caused by the lack of representativeness.*
> *Better:* "The data were collected using a self-administered questionnaire written in English; consequently the study sample omitted people with low literacy skills and those who do not speak English. In the United States, both of these groups are concentrated at the lower end of the socioeconomic spectrum; therefore estimates of health knowledge from this study probably overstate that in the general adult population." *This version clearly explains how the method of data collection affected who was included in the sample and the implications for estimated patterns of the outcomes under study. Jargon like "representativeness" and "target population" is restated in terms of familiar concepts, and the concepts are related to the particular research question and study design.*

Accompany your statements about limitations with reference to other publications that have evaluated those issues for other similar data. By drawing on others' work, you may be able to estimate the extent of bias without conducting your own analysis of each such issue.

Directions for Future Research

Close your discussion of limitations by listing some directions for future research. Identify ways to address the drawbacks of your study, perhaps by collecting additional data, or collecting or analyzing it differently, and mention new questions raised by your analyses. This approach demonstrates that you are aware of potential solutions to your study's limitations, and contributes to an understanding of where it fits in the body of research on the topic.

Box 10.3 is an example discussion of the advantages and disadvantages of the birth weight data, to follow the "results" material

BOX 10.3. DATA AND METHODS IN THE DISCUSSION OF
A SCIENTIFIC PAPER

"(1) This study of a large, nationally representative survey of US children extends previous research on determinants of low birth weight by including income and maternal smoking behavior—two variables not available on birth certificates, which were the principal data source for many past studies. (2) A potential drawback of the survey data is that information on birth weight was collected from the child's mother at the time of the survey—up to 10 years after the child's birth. In contrast to birth weight data from the birth certificate, which are recorded at the time of the birth, these data may suffer from retrospective recall bias. (3) However, previous studies of birth weight data collected from mothers reveal that they can accurately recall birth weight and other aspects of pregnancy and early infant health several years later (Githens et al. 1993; Olson et al. 1997). (4) In addition, racial and socioeconomic patterns of birth weight in our study are consistent with those based on birth certificate data (Martin et al. 2002), suggesting that the method of data collection did not appreciably affect results.

"(5) A useful extension of this analysis would be to investigate whether other Latino subgroups exhibit similar rates of low birth weight to those observed among the Mexican American infants studied here. (6) Inclusion of additional measures of socioeconomic status, acculturation, and health behaviors would provide insight into possible reasons for that pattern."

COMMENTS

(1) Identifies an advantage of the current data source over those used in previous analyses.

(2) Points out a potential source of bias in the study data.

(3) Cites previous research evaluating maternal retrospective recall of birth weight to suggest that such bias is likely to be small.

(4) Compares findings about sociodemographic determinants of birth weight from the survey with findings from studies using other sources of birth weight data to further dispel concerns

about the accuracy of the survey data, and provides citations to those studies.

(5) Rather than describing the inclusion of only one Latino group in the study sample as a weakness, suggests that expanding the range of Latino groups would be a useful direction for future research.

(6) Identifies additional variables that could shed light on the underlying reasons for the epidemiologic paradox (explained in box 11.3), again suggesting important ideas for later work.

presented in box 11.2b. By discussing possible strengths and weaknesses consecutively, the first paragraph helps to weigh their respective influences and provide a balanced assessment of the data quality.

CHECKLIST FOR WRITING ABOUT DATA AND METHODS

- Where possible, refer to other publications that contain details about the same data and methods.
- Regardless of audience, discuss how the strengths and limitations of your data and methods affect interpretation of your findings.

Consider audience and type of work to determine appropriate placement and detail about data and methods.

- For nonscientific audiences, see chapter 13.
- For scientific papers, reports, or grant proposals include a separate data and methods section.

    Use the W's to organize material on context and methods of data collection.

    Discuss the study response rate, issues of missing by design, loss to follow-up, and extent of item nonresponse on key variables

    Report the final analytic sample size in terms of number of cases and percentage of the original sample for which data were collected.

Define your variables.
- For familiar variables, define them as you describe the results.
- For novel or complex questions that are central to your analysis, include definitions or original wording in the text or an appendix.
- Report units, defining them if unusual or complicated.
- Describe how new variables were created, with a citation about that approach, if pertinent.
- Report on reliability and validity.

Discuss treatment of outliers and representativeness of sample.

Name the statistical methods.

Mention whether sampling weights were used, and if so, their source.

Describe strengths of the data and methods you used for addressing the topic of your study.

Discuss the limitations of your analysis, and suggest ways future research could address those limitations.

12345678910 11 1213

WRITING SCIENTIFIC PAPERS
AND REPORTS

The earlier material in this book laid the foundation for this chapter by introducing and demonstrating the principles and tools used to write about numbers on a variety of topics. In this chapter, I show how to put those building blocks together to craft a clear, well-integrated paper, report, or grant proposal on one topic. Academic papers or scientific reports about an application of a quantitative analysis usually follow a prescribed structure, with the text divided into an introduction, literature review, data and methods, results, and discussion and conclusions. Grant proposals follow a similar structure, substituting a description of pilot studies or preliminary findings for the results section, and replacing the discussion and conclusions with a section on policy, program, or research implications of the proposed project. Scientific reports typically replace an abstract with a one- to two-page executive summary (chapter 13), and include less detail and technical language in the literature review and data and methods. They summarize key findings in the text, accompanied by simplified tables and charts, relegating technical information about data sources, model specifications, and detailed tables of results to technical appendixes.

I begin this chapter with an overview of how to organize prose involving quantitative information, constructing a sequential, logical story with numbers used as evidence for the topic at hand. I then provide guidelines and examples of the structure and contents of introductory, results, and concluding sections before giving suggestions for writing a title, abstract, and keywords for a research paper on the relationships among race/ethnicity, socioeconomic status, and birth weight in the United States. See chapter 10 for how to write about data and methods. For additional guidance on writing scientific papers, see Montgomery (2003), Davis (1997), Hailman and Strier (1997), or Pyrczak and Bruce (2011). For guidelines on content, format, and style of materials for nonscientific audiences, see chapter 13.

ORGANIZING YOUR PROSE

Writing about numbers is similar to writing a legal argument. In the opening statement, a lawyer raises the major questions to be addressed during the trial and gives a general introduction to the characters and events in question. To build a compelling case, he then presents specific facts collected and analyzed using standard methods of inquiry. If innovative or unusual methods were used, he introduces experts to describe and justify those techniques. He presents individual facts, then ties them to other evidence to demonstrate patterns or themes. He may submit exhibits such as diagrams or physical evidence to supplement or clarify the facts. He cites previous cases that have established precedents and standards in the field, and discusses how they do or do not apply to the current case. Finally, in the closing argument he summarizes conclusions based on the complete body of evidence, restating the critical points but with far less detail than in the evidence portion of the trial.

Follow the same general structure as you write your quantitative story line for a scientific paper or grant proposal. The introduction parallels the opening argument; the data and methods and results sections mirror the evidence portion of the trial; and the discussion and conclusions parallel the closing argument. Open by introducing the overarching questions before describing the detailed numbers, just as a lawyer outlines the big picture before introducing specific testimony or other evidence to the jury. Describe and justify your methods of data collection and analysis. Systematically introduce and explain the numeric evidence in your exhibits—tables, charts, maps, or other diagrams—building a logical sequence of analyses. Close by summarizing your findings and connecting them back to the initial questions and previous studies of related topics.

As in other types of expository writing, introduce the broad topic or question of the work and then organize the factual material into separate paragraphs, each of which describes one major topic or pattern or a series of closely related patterns. Begin each paragraph with a statement of the issue or question to be addressed, write a sentence or two to sketch out the shape of the contrast or pattern with words, then provide and interpret numeric evidence to document that pattern (Miller 2006). To portray the sizes of each pattern, report results of selected comparisons such as rank, difference, ratio, or percentage difference between numeric values.

To describe a table or chart that encompasses more than one type of pattern, organize your narrative into paragraphs, each of which deals with one topic or set of closely related topics. For instance, a description of a chart showing trends in unemployment over two decades for each of several occupations might be organized into two paragraphs: the first describing trends over time and whether they are consistent for all the occupation categories, the second comparing levels of unemployment across occupational categories at one point in time and whether that pattern is consistent across time.

WRITING AN INTRODUCTION

In your introduction, state the topic and explain why it is of interest. After a general introductory sentence, report a few numbers to establish the importance of that topic. Include information on the prevalence of the issue or phenomenon under study or the consequences of that phenomenon, using one or two selected numeric contrasts to place those statistics in perspective. Provide a formal citation for each theory or numeric fact using the standard citation format for the journal or field for which you are writing; common formats include Chicago (University of Chicago 2010, used throughout this book), APA (American Psychological Association 2009), and AMA (American Medical Association 2007) styles. End the introduction with a statement of what your study will add to what is already known on the subject, either as a list of questions to be addressed or as one or more hypotheses.

Box 11.1 illustrates how to apply these ideas to the birth weight analysis, with comments keyed to numbered sentences within the narrative.

**BOX 11.1. USING NUMBERS IN AN INTRODUCTION
TO A SCIENTIFIC PAPER OR REPORT**

"(1) Low birth weight is a widely recognized risk factor for infant mortality, neurological problems, asthma, and a variety of developmental problems that can persist into childhood and even adulthood (US Environmental Protection Agency 2002; Institute of Medicine 1985). (2) For example, in 1999, US infants born weighing less than 2,500 grams (5.5 pounds) were 24 times as likely as normal birth weight infants to die before their first birthday (60.5 deaths per 1,000 live births and 2.5 deaths per 1,000, respectively; Mathews, MacDorman, and Menacker 2002). Although they comprised about 7.5% of all births, low birth weight infants accounted for more than 75% of infant deaths (Paneth 1995).

"(3) Costs associated with low birth weight are substantial: in 1995, Lewit and colleagues estimated that $4 billion—more than one-third of all expenditures on health care for infants—was spent on the incremental costs of medical care for low birth weight infants. Higher risks of special education, grade repetition, hospitalization, and other medical costs added more than $85,000 (in 1995 dollars) per low birth weight child to costs incurred by normal birth weight children through age 15 (Lewit et al. 1995).

"(4) Despite considerable efforts to reduce the incidence of low birth weight, the problem remains fairly intractable: between 1981 and 2000, the percentage of low birth weight infants rose from 6.8% to 7.6% of all infants, in part reflecting the increase in multiple births (Martin et al. 2002). (5) Rates of low birth weight among black infants remained approximately twice those among white infants over the same period (13.0% and 6.5% in 2000, respectively). (6) To what extent was that pattern due to the lower socioeconomic status of black children compared to children of other races in the United States? That question is the focus of this analysis."

COMMENTS

(1) Introduces the topic of the paper and gives a general sense of its importance, with reference to major studies on the topic.

(2) Reports statistics on the consequences of low birth weight, using relative risk and percentage share to quantify mortality differences, and citing original sources of the figures used in those comparisons.

(3) Reports estimates from other studies of the costs associated with low birth weight, providing further evidence that the topic merits additional study.

(4) Generalizes about trends in low birth weight over the two decades preceding the study, and reports numeric facts to illustrate those patterns, with citations of the original data sources.

(5) Presents information about racial differences in low birth weight, providing a transition to the research question for this study (sentence 6).

LITERATURE REVIEW

Academic papers, books, and grant proposals usually include a review of the existing literature on the topic to provide theoretical background on the hypothesized relationships among variables, and to report numeric findings for comparison with the current study. In some disciplines, the literature review is a freestanding section or chapter; in others, it is integrated into the introduction. If your literature review is separate from your introduction, place your research questions or hypotheses after the literature review because acquainting readers with theory and empirical findings from other studies helps substantiate the reasons behind the specific objectives and hypotheses for your analysis or proposed study.

Instead of writing separate sentences or paragraphs about each previous study, aim for a paragraph or two synthesizing previous findings on each of the major relationships you will examine. Summarize results of other studies, generalizing where possible about broad similarities and differences in theories or findings, and pointing out appreciable discrepancies among them. Where pertinent, discuss aspects of data or methods that affect interpretation or comparability of others' findings (see chapter 10 for issues to be considered).

As you discuss others' results, emphasize the direction and approximate size of associations, reporting a few illustrative numbers from those studies. Discuss only findings that are statistically significant, but mention when a lack of statistical significance for key variables contradicts theory or other studies (see "Statistical Significance" below). To facilitate comparison of results across studies, report the value as well as results of comparisons across values such as difference, ratio, or percentage change to provide the raw data for those calculations and to help readers interpret the contrasts.

To avoid a long, serial description of detailed findings from each of many studies, use the GEE ("generalization, example, exceptions") technique, explaining where there is and isn't consensus on the topic, and identifying questions that remain to be addressed. A dissertation or book can include more comprehensive discussion of individual studies, but should still provide a synthesis of current evidence and theory about the topic. See Pan (2008) for detailed guidelines and examples.

> *Poor:* "Smith and Jones (date) studied the relationship between race and birth weight in the United States and found [pattern *XYZ*]." Michaelson (date) also studied the relationship between race and birth weight and found [pattern *ABC*]. [Separate sentences describing results from each of five more studies on the topic.]
>
> *Better:* "Five out of seven recent studies (authors, dates) of the relationship between race and birth weight in the United States found that [pattern and example]. In contrast, Michaelson (date) found [different pattern and example], while DiDonato (date) found [yet another variant of the association among those variables] ..."

For literature reviews that address several issues, organize that material into cohesive subsections, each on one aspect of the topic. For example, if you are studying the relationship among two predictor variables and an outcome, organize the literature review into separate sections describing previous studies of how each of the predictors is associated with the outcome, and another subsection on the association between the two predictor variables. If your literature review on each of those subtopics spans more than a paragraph or two, use subheadings to orient readers to the subtopics. For the example topic we are tracing throughout this chapter, the corresponding subheadings of the literature review would read:

Race and Birth Weight
Socioeconomic Status and Birth Weight
Socioeconomic and Race

See the instructions for authors for the journal you are writing for to learn their conventions for formatting subheadings.

RESULTS SECTION

In the results section, report and describe numeric evidence to test your hypotheses, systematically presenting quantitative examples and contrasts. Organize the results section into paragraphs, each of which addresses one aspect of your research question. Start each paragraph with a sentence that introduces the topic of that paragraph and generalizes the patterns. Then present numeric evidence for those conclusions. A handful of numbers can be presented in a sentence or two. For more complex patterns, report the numbers in a chart or table, complemented with a description using the "generalization, example, exception" (GEE) approach. Refer to each table or chart by name as you describe the patterns and report numbers presented therein. As explained in chapter 10, you will have cited the source(s) of data used for your own analysis in your data and methods section, so you do not need to provide source citations for the facts in your results section.

Boxes 11.2a and 11.2b show "poor" and "better" descriptions of table 11.1, which comprises the first step in an analysis of whether socioeconomic or behavioral factors explain racial differences in birth weight. On the pages that follow, I critique the numbered elements from those descriptions, identifying the various principles for introducing, organizing, and describing statistical findings. (In the interest of space, I did not include a complete results section for this analysis, as the material shown below illustrates the relevant points.)

Table 11.1. Bivariate statistics on associations between key predictor and other independent variables

Birth weight, socioeconomic characteristics, and smoking behavior by race/ethnicity, United States, 1988–1994 NHANES III

	Non-Hispanic white (N = 3,733)	Mexican American (N = 3,112)	Non-Hispanic black (N = 2,968)	All racial/ethnic groups (N = 9,813)
Birth weight				
Mean (grams)	3,426.8	3,357.3	3,181.3	3,379.2
% Low birth weight[a]	5.8	7.0	11.3	6.8
Socioeconomic characteristics				
Mother's age				
Mean (years)	26.6	24.9	24.2	26.0
% Teen mother	9.4	18.4	22.9	12.5
Mother's education				
Mean (years)	13.3	9.1	11.9	12.6
% < High school	14.7	58.4	30.1	21.6
% = High school	34.9	24.5	41.7	35.0
Mean income-to-poverty ratio (IPR)[b]	2.60	1.34	1.39	2.28
% Poor	14.7	50.7	48.5	23.9
Health behaviors				
% Mother smoked while pregnant	26.8	10.1	22.9	24.5

Note: NHANES III = Third US National Health and Nutrition Examination Survey, 1988–1994. Statistics are weighted to population level using weights provided with the NHANES III (US DHHS 1997); sample size is unweighted. Differences across racial/ethnic origin groups were statistically significant for all variables shown (*p* < 0.01).

[a] Low birth weight < 2,500 grams or 5.5 pounds.

[b] The income-to-poverty ratio (IPR) is family income in dollars divided by the Federal Poverty Threshold for a family of comparable size and age composition. A family with income equal to the poverty threshold (e.g., $17,960 for two adults and two children in 2001; US Census Bureau 2002a) would have an income-to-poverty ratio of 1.0.

BOX 11.2A. DESCRIPTION OF A TABLE AND CHART FOR A RESULTS SECTION: POOR VERSION

"(1) (2) Race/ethnicity is strongly related to birth weight and LBW (both $p < 0.01$). Mean birth weight was 3,426.8, 3,181.3, and 3,357.3, for non-Hispanic white, non-Hispanic black, and Mexican American infants, respectively. (3) Average educational attainment, percentage of high school graduates, income, percentage poor, percentage of teen mothers, and mother's age are all statistically significant ($p < 0.01$). (4) An interesting finding is that smoking shows the opposite pattern of all the other variables ($p < 0.01$). (5) Figure 11.1 presents mean birth weight for each combination of race/ethnicity and (6) mother's educational attainment. (7) Among infants born to women with less than complete high school, mean birth weight was 3,300, 3,090, and 3,345 for non-Hispanic whites, non-Hispanic blacks, and Mexican Americans, respectively. [Sentences with corresponding numbers for each racial/ethnic group among =high school and college+]. (8) (9) (10)."

BOX 11.2B. DESCRIPTION OF A TABLE AND CHART FOR A RESULTS SECTION: BETTER VERSION

"(1) Table 11.1 presents weighted statistics on birth weight, socioeconomic characteristics, and maternal smoking for the three racial/ethnic groups, along with unweighted number of cases in each group. All differences shown are statistically significant at $p < 0.01$. (2) (A) On average, Non-Hispanic white newborns were 246 grams heavier than non-Hispanic black infants, and 70 grams heavier than Mexican American infants (3,427 grams, 3,181 grams, and 3,357 grams, respectively). The 176 gram birth weight advantage of Mexican American over non-Hispanic black infants was also statistically significant. (B) These gaps in mean birth weight translate into substantial differences in percentage low birth weight (LBW < 2,500 grams), with nearly twice the risk of LBW among non-Hispanic black as non-Hispanic white infants (11.3% and 5.8%, respectively). Mexican American infants were only slightly more likely than whites to be LBW (7.0%; relative risk = 1.2; $p < 0.05$)."

"(3) In every dimension of socioeconomic status studied here, non-Hispanic black and Mexican American mothers were substantially disadvantaged relative to their non-Hispanic white counterparts. They were twice as likely as non-Hispanic white mothers to be teenagers at the time they gave birth, and two to three times as likely to be high school dropouts. Mean income/needs ratios for non-Hispanic black and Mexican American families were roughly half that of non-Hispanic white families. (4) In contrast to the socioeconomic patterns, maternal smoking—an important behavioral risk factor for low birth weight—was more common among non-Hispanic white (27%) than non-Hispanic black (23%) or Mexican American women (10%).

"(5) Does the lower average socioeconomic status (SES) of non-Hispanic black and Mexican American infants explain their lower mean birth weight? To answer that question, figure 11.1 presents mean birth weight for the three racial/ethnic groups (6) within each of three socioeconomic strata, defined here in terms of mother's educational attainment. (7) Birth weight increased with increasing mother's education in each of the three racial/ethnic groups. In addition, at all mother's education levels, non-Hispanic black infants weighed 180 to 225 grams less than their non-Hispanic white or Mexican American counterparts (8). The birth weight deficits for non-Hispanic black compared to non-Hispanic white infants within each education level were statistically significant at $p < 0.05$, as illustrated by the fact that the 95% confidence intervals around their respective mean birth weights do not overlap one another. (9) Among infants born to mothers with at least some college, non-Hispanic white infants outweighed Mexican Americans ($p < 0.05$). In the lowest two mother's education groups, on the other hand, Mexican American infants slightly outweighed their non-Hispanic white peers, although those differences were not statistically significant.

"(10) These statistics show that even within socioeconomic strata defined by mother's educational attainment, non-Hispanic black race is associated with substantially lower birth weight than the other racial/ethnic groups."

Statement 1

Poor: [No introductory sentence.]

By jumping right into a description of the table, this version fails to orient readers to the purpose of that table, which is not named.

Better: "Table 11.1 presents weighted statistics on birth weight, socioeconomic characteristics, and maternal smoking for the three racial/ethnic groups, along with unweighted number of cases in each group. All differences shown are statistically significant at $p < 0.01$."

The topic sentence names the associated table and introduces its purpose, restating the title into a full sentence. It also specifies which statistics are weighted and which are unweighted. The second sentence generalizes about statistical significance for the entire table, echoing the footnote to the table and eliminating the need to report results of statistical tests for each association.

Statement 2

Poor: "Race/ethnicity is strongly related to birth weight and LBW (both $p < 0.01$). Mean birth weight was 3,426.8, 3,181.3 and 3,357.3, for non-Hispanic white, non-Hispanic black, and Mexican American infants, respectively."

This sentence reports but does not interpret average birth weight for each racial/ethnic group, adding little to the information in the table. By not referring to the associated table by name, it makes it harder for readers to see the numeric basis of the author's conclusions. It omits the units in which birth weight is reported, and presents a string of numbers that is visually difficult to take in. It's always good to separate large numbers with a little text.

Better: "(A) On average, non-Hispanic white newborns were 246 grams heavier than non-Hispanic black infants, and 70 grams heavier than Mexican American infants (3,427 grams, 3,181 grams, and 3,357 grams, respectively). The 176 gram birth weight advantage of Mexican American over non-Hispanic black infants was also statistically significant. (B) These gaps in mean birth weight translate into substantial differences in percentage low birth weight (LBW < 2,500 grams), with nearly twice the risk of LBW among non-Hispanic black as non-Hispanic white infants (11.3% and 5.8%, respectively). Mexican American infants were only slightly more likely than whites to be LBW (7.0%; relative risk = 1.2; $p < 0.05$)."

The first and second sentences (A) report direction and size of differences in mean birth weight across racial/ethnic groups using difference (subtraction) to assess size, and reporting the units of measurement. The last two sentences (B) quantify differences in the prevalence of low birth weight across racial/ ethnic groups using relative risk (division), and report the rates used in those calculations. The LBW acronym and definition of low birth weight are repeated from the methods section to remind readers of their meaning.

Statement 3

Poor: "Average educational attainment, percentage of high school graduates, income, percentage poor, percentage of teen mothers, and mother's age are all statistically significant ($p < 0.01$)."
This seemingly simple sentence is plagued by numerous problems.

- It does not explain that the statistical tests are for differences across racial/ethnic groups in each of the socioeconomic status (SES) variables. Because race is not mentioned, readers may mistakenly think that the tests are for association among the SES variables. Combined with the absence of an introductory sentence, this lack of explanation leaves the results almost completely disconnected from the research question.
- Both the continuous and categorical versions of each SES measure are reported without pointing out that they are merely two different perspectives on the same concept. For example, "percentage poor" simply classifies family income into poor and non-poor.
- The variables are named in a different order in the table and text.

Better: "In every dimension of socioeconomic status studied here, non-Hispanic black and Mexican American mothers were substantially disadvantaged relative to their non-Hispanic white counterparts. They were twice as likely as non-Hispanic white mothers to be teenagers at the time they gave birth, and two to three times as likely to be high school dropouts. Mean income/needs ratios for non-Hispanic black and Mexican American families were roughly half that of non-Hispanic white families."

Because most of the remaining numbers in table 11.1 deal with associations between race/ethnicity and socioeconomic status, a new paragraph is started to describe those findings. The topic sentence introduces the concepts to be discussed and generalizes the patterns. The next two sentences illustrate that generalization with specific comparisons from the table, using ratios to quantify racial/ethnic disparities in the three socioeconomic measures.

Statement 4

Poor: "An interesting finding is that smoking shows the opposite pattern of all the other variables ($p < 0.01$)."

"Opposite" of what? This sentence does not indicate which patterns smoking is being compared against or mention the direction of any of the patterns. Neither this nor the preceding sentence (statement 3) mentions that the associations are with race/ethnicity, again keeping the description divorced from the main purpose of the analysis.

Better: "In contrast to the socioeconomic patterns, maternal smoking—an important behavioral risk factor for low birth weight—was more common among non-Hispanic white (27%) than non-Hispanic black (23%) or Mexican American women (10%)."

This version points out an exception to the generalization that people of color are worse off, reporting the higher smoking rates among non-Hispanic whites.

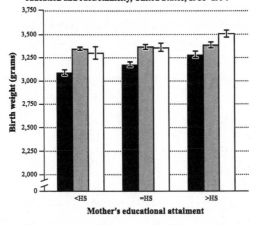

Mean birth weight and 95% confidence intervals, by mother's education and race/ethnicity, United States, 1988–1994

Non-Hispanic black   Mexican American   Non-Hispanic white

Figure 11.1. Clustered bar chart of three-way association
Source: Miller 2013a, table 15.2.
Note: Data are from the Third US National Health and Nutrition Examination Survey (US DHHS 1997). Weighted to national level using sampling weights provided with the NHANES III (US DHHS 1997).

Statements 5 and 6

Poor: "Figure 11.1 presents mean birth weight for each combination of race/ethnicity and mother's educational attainment."
This version states the topic but not the purpose of the figure. By omitting a transition from the pattern in table 11.1 to the pattern in figure 11.1, this description leaves readers without a sense of how the evidence in that table and figure fits together, or how those analyses relate to the overall research question.

Better: "(5) Does the lower average socioeconomic status (SES) of non-Hispanic black and Mexican American infants explain their lower mean birth weight? To answer that question, figure 11.1 presents mean birth weight for the three racial/ethnic groups (6) within each of three socioeconomic strata, defined here in terms of mother's educational attainment."
The sentences in statement 5 orient readers to the purpose of the different steps in the analysis, summarizing the findings in table 11.1 and explaining how figure 11.1 builds upon those findings.

*By starting a new paragraph, this version provides a segue from
the table—which documented the racial differences in both
birth weight and socioeconomic status—to a figure showing
the three-way association among race/ethnicity, SES, and mean
birth weight. Statement 6 explains that mother's education is a
measure of SES, helping orient readers to the purpose of this step
in the analysis.*

Statement 7

Poor: "Among infants born to women with less than complete
high school, mean birth weight was 3,300, 3,090, and 3,345
for non-Hispanic whites, non-Hispanic blacks, and Mexican
Americans, respectively. [Sentences with corresponding
numbers for each racial/ethnic group in the high school
diploma and college+ groups.]"
*By simply listing nine birth weight values, this version fails
to portray the size or shape of the two general patterns (birth
weight by race and education), or relate them to the underlying
research question. It is a classic example of reporting but not
interpreting numeric evidence.*

Better: "Birth weight increased with increasing mother's
education in each of the three racial/ethnic groups. In addition,
at all mother's education levels, non-Hispanic black infants
weighed 180 to 225 grams less than their non-Hispanic white
or Mexican American counterparts."
*This version describes the birth weight patterns by race/ethnicity
and educational attainment using the GEE approach. By
comparing non-Hispanic black infants to the other two racial/
ethnic groups, and describing the direction and magnitude
of that difference, it ties the evidence back to the underlying
research question.*

Statement 8

Poor: [No mention of statistical significance of birth weight
differences across racial/ethnic or education groups.]
Better: "The birth weight deficits for non-Hispanic black compared
to non-Hispanic white infants within each education level

were statistically significant at $p < 0.05$, as illustrated by the fact that the 95% confidence intervals around their respective mean birth weights do not overlap one another."

This sentence explains the direction and statistical significance of the racial differences in birth weight, with a brief explanation of how to interpret the confidence intervals shown in the figure.

Statement 9

Poor: [No discussion of specific contrasts of racial/ethnic groups within education levels, or vice versa.]

Better: "Among infants born to mothers with at least some college, non-Hispanic white infants outweighed Mexican Americans ($p < 0.05$). In the lowest two mother's education groups, on the other hand, Mexican American infants slightly outweighed their non-Hispanic white peers, although those differences were not statistically significant."

These sentences describe how the direction of the birth weight pattern for Mexican American compared to non-Hispanic white infants varies by education level, demonstrating that that pattern is different from that comparing non-Hispanic black infants to either group. Again, direction, magnitude, and statistical significance are all discussed.

Statement 10

Poor: [No summary of the shape of the association among race, mother's education, and birth weight].

By omitting a summary of the findings of table 11.1 or figure 11.1, this description doesn't help readers see what those analyses demonstrated about the overall research question.

Better: "These statistics show that even within socioeconomic strata defined by mother's educational attainment, non-Hispanic black race is associated with substantially lower birth weight than the other racial/ethnic groups."

This narrative provides a clear summary of the conclusions about race, mother's education, and birth weight, tying it back to the original research question.

DISCUSSION AND CONCLUSIONS

In the discussion and conclusions section, relate the evidence from your analysis back to the larger research question, comparing your main findings against hypotheses posed at the beginning of the work and results from related studies, with appropriate citations. See also chapter 10 regarding data and methods in the conclusions.

Numeric Information in a Concluding Section

Restate conclusions about the size and statistical significance of associations among the variables in the main research question, and consider extensions that help readers see the importance (or lack of importance) of those findings. To convey the purpose and interpretation of numeric facts or contrasts, introduce them in sentences that place them in their substantive context.

EFFECT SIZE

Rather than repeat precise numeric estimates and results of statistical tests from the results section, use verbal descriptions or approximate numeric values, rounding to the nearest whole multiple or familiar fraction.

STATISTICAL SIGNIFICANCE

Rarely is statistical significance discussed explicitly in the concluding section, and then only for key variables in your research question. Use results of statistical tests in conjunction with substantive considerations to assess which findings to emphasize. Instead of reporting standard errors, test statistics, *p*-values, or confidence intervals in the discussion section, use phrases such as "were statistically significant" or "was not associated." See "Discussion Sections" in chapter 3 for examples.

There are three situations where you should discuss statistical significance in your concluding section:

(1) If your statistical test results *run counter to theoretical expectations*, such as when theory led you to predict a large, statistically significant difference across groups that was not borne out in your study, or vice versa.

(2) If your statistical test results *conflict with those of previous empirical studies.* Perhaps you found statistically significant differences that others had not. Or others may have found statistically significant differences that were not apparent with your data.

In those instances, explicitly mention the discrepant findings regarding statistical significance, using words rather than detailed numeric results. Explain what these findings imply, relating them back to your original hypotheses and the literature that led you to formulate those hypotheses. Discuss possible explanations for the discrepancy of findings across studies, such as differences in study populations, design, or variables.

(3) The third situation in which statistical significance merits discussion is if you observe changes in effect size and statistical significance of key variables in your model when you introduce measures of potentially mediating or confounding factors, particularly those that were not previously available or were poorly measured in other studies. This kind of issue is often the reason for estimating a multivariate model (Miller 2013a).

CAUSALITY AND SUBSTANTIVE SIGNIFICANCE REVISITED

To bring your analysis to a close, describe the implications of the associations reported in the results section. In analyses that ask cause-and-effect type questions, revisit two issues discussed in chapter 3: First, can the associations be viewed as causal? And second, if so, what is the substantive meaning or importance of the findings? As you describe the relationships in your analysis, choose wording that conveys whether you are interpreting those relationships as causal or merely as correlated phenomena. To assess how much a pattern matters, combine estimates from the analytic portion of the paper with information from other sources. Depending on your topic, these calculations might involve cost effectiveness analysis (e.g., Gold et al. 1996), attributable risk calculations (chapter 5), or other applicable measures of net benefits, costs, or tradeoffs between alternative proposed solutions.

Citing Other Sources

Unlike the results section, which is devoted almost exclusively to reporting and interpreting data from within your own study, a discussion and conclusions section often incorporates numeric information from other studies. There are several reasons to cite other works in the discussion:

- To evaluate whether the current findings are consistent with previous research
- To provide perspective on the importance of the topic
- To apply numbers from other sources of background information that place the findings of the current study in context
- To compare your results with those obtained using different analytic methods or study designs
- To offer ideas about underlying reasons for observed patterns, generating hypotheses to be tested empirically in later studies (otherwise known as "directions for future research")

As you restate your own findings and compare them against others' theories or results, using phrasing to differentiate between numeric information that comes from your own analyses and that from other sources. Doing so is essential in the discussion section because many paragraphs in that section interpret the results of your studies in the broader context of what is known on your topic from previous studies. Box 11.3 is an illustrative discussion and conclusions section of a scientific paper on racial/ethnic differences in birth weight, summarizing the findings presented in the results section (box 11.2b above).

TITLE

The title is the first aspect of your paper or proposal readers will see. Use it to identify the topic, context, and what is new or different about your work—what sets it apart from other related studies. Briefly convey the main questions or hypotheses you investigate, using your research question and context (W's) as starting points. Often, a rhetorical version of your research question or objective works well.

"(1) Consistent with a large body of previous research (e.g., Institute of Medicine 1985), we found a substantial birth weight disadvantage among non-Hispanic black infants compared to infants from other racial/ethnic groups in the United States. (2) Although non-Hispanic black infants have on average twice the incidence of low birth weight of non-Hispanic white infants, (3) in large part this difference can be attributed to the fact that black infants are more likely to be of low socioeconomic status (SES). Regardless of race, children born into low SES families have lower mean birth weight than those born at higher SES. (4) However, differences in family socioeconomic background do not explain the entire difference across racial/ethnic groups. At each socioeconomic level, non-Hispanic black newborns weigh on average considerably less than their non-Hispanic white counterparts, with particularly marked gaps at higher socioeconomic levels. These differences in mean birth weight generate a widening in risk of low birth weight (LBW < 2,500 grams) for black infants as SES rises: for infants born to mothers with less than complete high school education, LBW is 30% more common among black infants than among white infants; for those born to mothers who attended college, the excess risk for blacks is 125%.

"(5) The causal role of low socioeconomic status is also brought into question by the relatively low incidence of LBW among Mexican American infants, which is quite close to that of non-Hispanic white infants and is far below that of non-Hispanic black infants. This phenomenon of relatively good health among Mexican Americans despite their low SES is referred to as the "epidemiological paradox" or "Hispanic paradox" and also has been observed for other health conditions (Franzini et al. 2001; Palloni and Morenoff 2001).

"(6) Other possible mechanisms that have been proposed to explain why black infants are more likely to be low birth weight include less access to health care, higher rates of poor health behaviors, greater social stress (Zambrana et al. 1999), intergenerational

BOX 11.3. (CONTINUED)

transmission of health disadvantage (Conley and Bennett 2000), and other unmeasured factors that affect black people more than those of other racial/ethnic origins.

"(7) Reducing the incidence of low birth weight is a key objective of Healthy People 2010 (US DHHS 2001). (8) If the incidence of low birth weight among black infants had been decreased to the level among white infants, nearly 40,000 of the low birth weight black infants born in 2000 would instead have been born at normal birth weight. That reduction in low birth weight would have cut the black infant mortality rate by more than one-third, assuming the infant mortality rate for normal birth weight infants (Mathews, Mac-Dorman, and Menacker 2002). In addition, an estimated $3.4 million in medical and educational expenses would be saved from that birth cohort alone, based on Lewit and colleagues' estimates of the cost of low birth weight (1995)."

COMMENTS

(1) Generalizes the major finding of the current study and places it in the context of previous research, citing a summary report by a prominent national research institute. Use of first person language ("we found") clearly identifies which findings were from the current study.

(2) Quantifies the size of the black/white difference in low birth weight with approximate figures ("twice the incidence") rather than reporting exact percentages for each group.

(3–6) Describes the association between race/ethnicity in cause-neutral language ("found a substantial birth weight disadvantage among," sentence 1), leading into a discussion of causal interpretation. In contrast, the hypothesized causal role of socioeconomic characteristics is clearly conveyed using language such as "attributed to" (sentence 3) and "do not explain" (sentence 4). The intentional use of causal language continues into the subsequent paragraph with "causal role" (sentence 5), and "possible mechanisms" (sentence 6).

Sentences 5 and 6 discuss possible explanatory pathways linking race and low birth weight, citing published sources of these theories. Sentence 5 mentions the "epidemiological paradox" observed in other health studies and relates it to the current findings. Sentence 6 introduces other theories that can be used to generate hypotheses to be tested in future studies. Ideas derived from other sources are differentiated from results of the current study using phrases such as "has been observed for other health conditions," "Other possible mechanisms that have been proposed," and "[Author] and colleagues' estimates." The specific sources of each theory or fact are identified by formal citations.

(7) Brings the paper full circle, returning to the "big picture" to remind readers of the reasons for addressing this research question. Establishes that lowering the incidence of low birth weight is a major priority identified by experts in the field, citing the pertinent policy document.

(8) Combines statistics on the excess risk of low birth weight among black infants from the current analysis with information from other published sources about infant mortality rates and costs of low birth weight to estimate how many infant deaths could be prevented and how many dollars saved if the incidence of low birth weight among black infants could be reduced to the same level as that among whites. Again, figures are reported in round numbers: phrases such as "nearly 40,000" and "more than one-third" are precise enough to make the point.

- If you are analyzing a relationship among two or three concepts, use the title to name those concepts or to state your hypothesis about how they are related. For instance, "The Contribution of Expanding Portion Sizes to the US Obesity Epidemic" (Young and Nestle 2002) clearly identifies the predictor and outcome variables—expanding portion sizes and obesity, respectively. Reworded as a rhetorical question, the title might read: "The US Obesity Epidemic: How Much Did Expanding Portion Sizes Contribute?"

- If you are describing a trend or other pattern, name the relevant dimensions of the contrast in the title. For example, "Voter Turnout from 1945 to 1998: A Global Participation Report" (International Institute for Democracy and Electoral Assistance 1999) indicates both the period to which the analysis pertains and the international scope of the comparison, as well as the topic under study.
- If your work is among the first to advance a new type of study design or statistical method, or to apply a technique to a new topic, mention both the method and the topic in the title. For example, "Dropping Out of Advanced Mathematics: A Multilevel Analysis of School- and Student-Level Factors" captures the outcome variable, statistical method, and different levels of analysis for the predictor variables. For a lay audience, replace the name of the method with the question it is intended to answer, such as "Dropping Out of Advanced Mathematics: How Much Do Students and Schools Contribute to the Problem?" (Ma and Willms 1999).

Titles are often best drafted early in process of writing a paper to help you establish the main focus of the paper, then tweaked when the paper has been completed to make sure the title effectively captures the essence of the project.

ABSTRACTS AND KEYWORDS

Most scientific papers or proposals require an abstract and keywords to summarize and classify the work. Conference organizers often review abstracts rather than full papers when deciding among submitted papers.

Abstracts

Abstracts are capsule summaries of a scientific paper, grant proposal, research poster, or book. They are often shown on the first page of the associated journal article or research proposal, in the program book for a professional conference, or in a list of recently published books. They provide a quick way for readers to familiarize themselves with the topic and findings of a large number of studies, often helping them to decide which articles to read closely, which presentations to attend, or which proposals to review in detail. Abstracts are your opportunity to stand out among competing works, so take the time to write a compelling, accurate summary of your work.

The length and format of abstracts vary by discipline and type of publication, but the contents generally include short descriptions of the objectives, data and methods, results, and conclusions of the study. Structured abstracts include subheadings for each of those parts, while unstructured abstracts integrate all those elements into one paragraph. Consult the instructions for authors for your intended publication for guidelines on structure and length.

Use the W's to organize the information in an abstract.

- In the objectives (sometimes called the "purpose" or "background") section, state what your study is about and, if word count permits, why it is important.
- In the data and methods part of the abstract (sometimes divided into "data sources," "study design," and "data collection and analysis" or other similarly-named subsections), state the type of study design, listing who, what, when, where, and how for each major data source, mentioning the number of cases, and naming the analytic methods.
- In the results section of the abstract, briefly summarize key findings, including a few carefully chosen numbers and statistical test results.
- In the conclusions (sometimes called "discussion," or divided into "conclusions" and "policy implications"), relate major findings to the initial objectives. For journal that aims for readers interested in applications of the findings as well as research-oriented audiences, include a short phrase or two pointing out policy, practice, or program implications of the findings.

Although the abstract will be printed above the body of the work, I recommend that it be written last, because it must capture elements from each section of the paper, proposal, or presentation. Alternatively, draft it as you write the paper to create placeholders for the key information, and then revise the abstract once all analyses and the body of the paper have been finalized to make sure it reflects the final numbers and conclusions.

Keywords

Keywords are used in online databases of publications and conference presentations, allowing readers to search for papers that match a specified set of topics, methods, or contextual characteristics. As the term suggests, keywords should emphasize key elements of the paper; use the W's as a mental checklist to identify them. At a minimum, include keywords for the outcome and key predictor variables. Also include location or dates if these are important aspects of your research question, and mention any major restrictions on who was studied.

Box 11.4 includes a title, structured 150-word abstract, and keywords for the analysis of racial differences in birth weight, with numbered comments for each component.

BOX 11.4. TITLE, STRUCTURED ABSTRACT, AND KEYWORDS FOR A SCIENTIFIC PAPER

Title: Does Socioeconomic Status Explain Racial/Ethnic Differences in Birth Weight in the United States?

Objectives. (1) To assess whether differences in socioeconomic status explain lower mean birth weight among non-Hispanic black compared to infants of other race/ethnicity in the United States.

Methods. (2) *t*-tests for differences in means were used to analyze (3) data from the Third National Health and Nutrition Examination Survey (NHANES III, 1988–1994) for 9,813 children. (4) Birth weight and socioeconomic characteristics were from the parental interview.

Results. (5) Even within socioeconomic strata defined by mother's

---

education, (6) non-Hispanic black infants weighed 180 to 225 grams less than non-Hispanic whites or Mexican Americans ($p < 0.01$). (7) Mexican Americans weighed less than non-Hispanic whites only in the mother's highest education group ($p < 0.05$).

Conclusions. (8) At all socioeconomic levels, non-Hispanic black infants weighed considerably less than Mexican American or non-Hispanic white infants. (9) Additional research is needed to identify modifiable risk factors to improve birth weight among black infants.

Keywords: (10) birth weight; black race; Hispanic ethnicity; socioeconomic factors; United States.

COMMENTS

(1) Identifies the predictor and outcome variables and sets the geographic context.
(2) Specifies the statistical methods.
(3) Names the data source, including dates, age group, and number of cases.
(4) Reports how the variables were collected.
(5) Describes the analytic method used to control for socioeconomic status.
(6) Summarizes the direction, magnitude, and statistical significance of differences in mean birth weight for non-Hispanic black compared to other infants.
(7) Describes the shape of the association between education and birth weight for Mexican American versus non-Hispanic white infants.
(8) Restates the overall findings of the analysis, tying them back to the research question (objective).
(9) Suggests an additional research direction and relates it to the underlying policy question ("modifiable risk factors to improve birth weight").
(10) Lists five keywords that identify the predictor and outcome variables, and context.

CHECKLIST FOR WRITING SCIENTIFIC PAPERS AND REPORTS

- Apply the principles for good expository writing.
  Organize ideas into one major topic per paragraph.
  Start each with an introductory sentence that identifies the purpose of that paragraph.
  Write transition sentences to show how consecutive paragraphs relate to one another.
- In the introduction,
  introduce the issues to be discussed,
  provide evidence about their importance, and
  state hypotheses; if the literature review is a separate section, put hypothesis after that section.
- In the literature review,
  organize paragraphs by major topic, grouping articles on each topic, rather than writing separately about every article;
  use subheadings to organize material on subtopics; and
  summarize theory and findings from previous studies of your topics, using the GEE approach to identify similarities and differences.
- In the results,
  mention context (W's);
  specify units and categories for all variables;
  keep jargon to a minimum, defining it when needed;
  specify the direction, magnitude, and statistical significance of associations;
  use dimensions of quantitative comparison to interpret numbers (chapter 5);
  synthesize patterns rather than repeating all of the detailed numbers from associated tables and charts;
  refer to charts and tables by name as you discuss them.
- In the discussion and conclusions,
  relate your findings back to your original hypotheses and to previous studies, summarizing key numeric findings rather than reporting detailed statistics,
  illustrate substantive significance of findings (see chapter 3),

discuss advantages and limitations of your data and analytic
approach (see chapter 10), and

identify implications for policy or research.

- In the title, mention your main topic and key facets of
context (W's).
- In the abstract,

summarize the objectives, methods, results, and conclusions of
your study;

use the W's to organize the topics; and

see instructions to authors for guidelines about length and
structure of abstract for the pertinent journal or call for
proposals.

- For the keywords, select nouns or phrases to identify the
predictor and outcome variables and context of your study.

1 2 3 4 5 6 7 8 9 10 11 **12** 13

SPEAKING ABOUT NUMBERS

Speaking about numbers is a common means of communicating quantitative information, whether a lecture in a classroom setting, a short speech to the general public, or a professional conference presentation. Many of the principles described throughout this book apply to speaking about numbers. However, there are a few important modifications that will improve your speeches about quantitative concepts or help you translate written documents into spoken form. The first section of this chapter includes a quick overview of time and pacing, use of visual materials, and speaker's notes, with an emphasis on aspects of public speaking that pertain specifically to conveying quantitative information. The second section describes how to create slides for a speech, including text, tabular, and graphical slides and their applications. The third section explains how to write speaker's notes to accompany your slides, including my infamous "Vanna White" technique for succinctly but systematically describing a table or chart. The last section provides guidance on rehearsing your speech to make sure it is clear and fits within the allotted time. See also Kosslyn (2007) and Briscoe (1996) for guidance on preparing slides, Montgomery (2003) or Hailman and Strier (1997) for suggestions on speaking to scientific audiences, and Nelson et al. (2002) for recommendations on speaking to applied audiences.

CONSIDERATIONS FOR PUBLIC SPEAKING

Four factors together determine how you will design and deliver a speech: your topic, methods, audience, and the time available to you. Leave out any of those elements as you plan and your talk will not be as successful. For example, the appropriate depth, pace, types of materials, and language for describing results of an analysis of the relationship between exercise, diet, and obesity are very different for a five-minute presentation to your child's fifth-grade class than for a 10-minute talk to a school board nutrition committee, or a half-hour presentation to a panel of experts at the National Institutes of Health.

First identify the few key points you want your listeners to understand and remember, taking into account your topic, methods, and audience. Then consider time and pacing before you design the visual materials and speaker's notes.

Time and Pacing

Most speeches have been allocated a specific amount of time, whether five minutes, 15 minutes, or an hour or longer. There are tradeoffs between the length of time, the amount of material, and the pace at which you must speak. Reduce the range and depth of coverage rather than speeding up your delivery, especially for an audience that is not accustomed to quantitative information. Better to cut detail than to rush an explanation of your central points or fail to leave time for questions and discussion.

Although each person reads a written document at his own pace, members of your listening audience all receive the material at the same rate—the pace at which you show the slides and explain them. During a speech, individuals cannot take extra time to examine a chart or table, or go back to reread an earlier point. Set the tempo to meet the needs of your typical listener rather than aiming to please either the least or most sophisticated members of your audience. Even for scientific audiences, avoid moving at too rapid a clip. If you present results of many different analyses in a short talk, the findings blur together and the purpose of each gets lost. Decide which results relate to your main objectives, then introduce and explain them accordingly.

Visual Accompaniment

For speeches of more than a few minutes, visual materials focus your audience's attention and provide a structure to your speech. Slides also help listeners recall facts or concepts to which you refer. In the absence of visual reminders, spoken numbers are easily forgotten, so if specific values are important, put them on a slide. This point is doubly true for comparisons, patterns, or calculations: even if you elect not to create slides for every facet of your talk, do provide charts and tables for your audience to follow as you describe key patterns or findings so they don't have to try to envision them as you speak.

A complete set of slides guides you through your material in a logical order and reminds you where you were if you stopped to answer questions from the audience. Some speakers like to create slides for each component of their talks, mixing text and picture slides for introductory, background, and concluding material with charts and tables of numeric results. However, some speakers prefer a less formal approach, with slides only of essential tables and charts. Even if you use a comprehensive set of slides in some situations, you may want only selected slides in others. For example, although I usually create slides for the whole talk for short professional presentations, I rarely use that approach when teaching. I've found that putting every aspect of a lecture on slides discourages student participation, so I generally create slides only of tables, charts, or other diagrams that I plan to discuss. Working from a written outline or notes, I then introduce each topic, interweaving questions that require students to supply details from readings, describe patterns in the charts or tables, practice calculations, or provide illustrative anecdotes for the points under discussion.

To decide among these different approaches, consider the available time and your own experience, style, and desired extent of interaction with your audience.

Speaker's Notes

Effective slides reduce full sentences into short phrases and reduce complex tables and charts into simpler versions. Accompanying speaker's notes include full sentences and paragraphs to introduce, flesh out, and summarize the information on each slide, and to provide the wording of transitions between slides. For a "generalization, example, exceptions" (GEE) description of a chart or table, speaker's notes are a place to store clear, concise, well-organized explanations that you have tested on similar audiences. Notes can prompt you about which aspects of tables or graphs to emphasize, or remind you of good examples, analogies, or anecdotes to reinforce points on the slide. Perhaps most important, speaker's notes are a reminder *not* to simply read the material on your slide out loud—a truly deadening way to give a presentation. More detailed guidelines on writing speaker's notes are given below.

SLIDES TO ACCOMPANY A SPEECH

Slides focus and direct your audience and display the facts and patterns mentioned in the speech. Presentation software such as PowerPoint makes it easy to produce text, tabular, and graphical slides, and accompanying speaker's notes. Such software automatically formats the material with large type, bullets, and other features that enhance readability and organization. Once the slides have been created, it is simple to reorganize text within and across slides, adding or removing material to create longer or shorter versions of talks, or making other revisions. Depending upon available audiovisual and computing equipment, these materials can be projected from a computer directly onto an auditorium screen, shared electronically with your audience members to view and notate on their own devices as PDFs or slides, or printed as paper handouts.

Some years ago, a backlash emerged against the use of PowerPoint and other presentation software, stating that these programs have led to inferior content and organization of slides, overreliance on fancy graphics, and substitution of rote reading of slides for other, more engaging means of presentation (Schwartz 2003; Tufte 2003). Used poorly, any tool—whether a hammer, paintbrush, or presentation software—can be used to produce shoddy work. With appropriate training and good technique, however, these tools can help create exemplary results. Below are guidelines on how to create effective slides for a speech, whether or not you elect to use presentation software.

Organizing Your Talk

For a speech to an academic audience, organize your talk with sections that parallel the sections of a scientific paper: an overview and introduction, review of the key literature, description of your data and methods, results, and conclusions. Below are illustrative slides for the sections of a scientific talk about racial/ethnic and socioeconomic differences in low birth weight based on the material in chapters 10 and 11. These slides can also be used as the basis for a research poster, or modified to create a chartbook about your findings. For a talk to an applied audience, devote less time to previous literature or data and methods, focusing instead on the purpose, results, and conclusions of your study. See chapter 13 for more suggestions about posters, and chartbooks, appendix B for a comparison of format and content for papers, speeches, and posters about a research project.

INTRODUCTION, OVERVIEW, AND LITERATURE REVIEW

In the introduction, familiarize your audience with your topic: what are the main issues you will be investigating, and why are they interesting and important? Incorporate some background statistics about the consequences of the issue under study (figure 12.1) or provide some statistics on the frequency with which it occurs (figure 12.2).

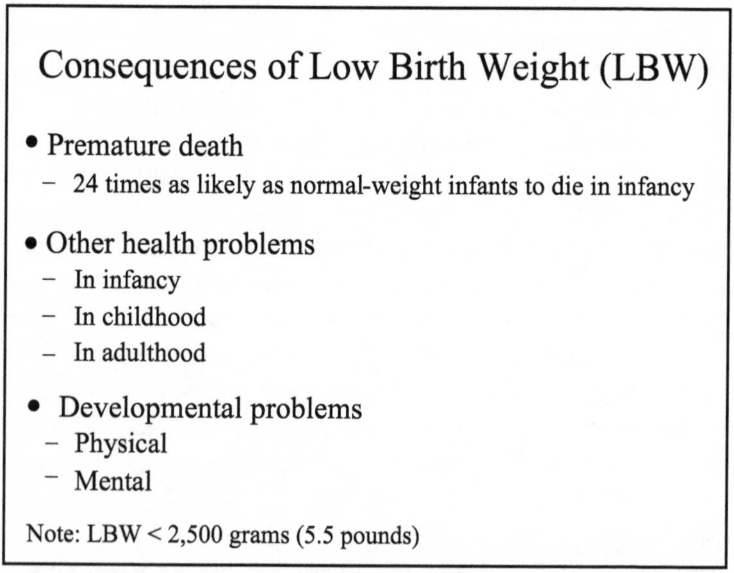

Consequences of Low Birth Weight (LBW)

- Premature death
  - 24 times as likely as normal-weight infants to die in infancy

- Other health problems
  - In infancy
  - In childhood
  - In adulthood

- Developmental problems
  - Physical
  - Mental

Note: LBW < 2,500 grams (5.5 pounds)

Figure 12.1. Introductory slide: Bulleted text on consequences of issue under study
Sources: Martin et al. 2002; US Environmental Protection Agency 2002; Institute of Medicine 1985

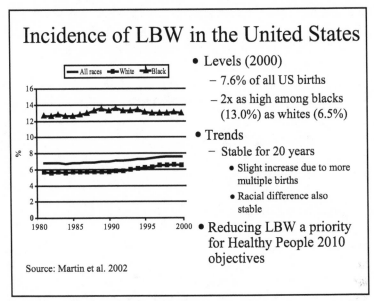

Incidence of LBW in the United States

— All races ■ White ▲ Black

- Levels (2000)
  - 7.6% of all US births
  - 2x as high among blacks (13.0%) as whites (6.5%)
- Trends
  - Stable for 20 years
    - Slight increase due to more multiple births
    - Racial difference also stable
- Reducing LBW a priority for Healthy People 2010 objectives

Source: Martin et al. 2002

Figure 12.2. Introductory slide: Chart and text on incidence of issue under study

For speeches of 20 minutes or more, consider starting with an overview slide that outlines the topics you will cover (figure 12.3).

Unless you have half an hour or more for your speech, devote much less attention to reviewing the published literature on your topic than you would in an academic paper of the same study. Often you can incorporate a few essential citations into your introduction. If a comparison of individual articles is important, consider summarizing their key conclusions on your topic in tabular form (e.g., figure 12.4).

DATA AND METHODS

Introduce your data, starting with the W's (who, what, when, where, and how many), type of study design, and response rates for your data sources (figure 12.5), and key variables (figure 12.6). Break this material up onto several slides to make room for all those topics without creating a cluttered slide. Define your variables on one or more slides in the data and methods section. If you define them as you present the results, viewers tend to focus on the numeric findings rather than listening to how the variables were measured and defined. For

Overview

- Low birth weight defined
- Importance
- Trends
- Data
- Results
- Conclusions

Figure 12.3. Slide outlining contents of speech

Previous Studies of Race & Birth Weight

Article	Type of study & data source	RR of LBW: black/white	Comments
Smith & Jones (1999)	Sample survey; birth certificates	2.2*	Nationally representative; controlled education
Williams (2000)	Retrospective survey; maternal questionnaires	3.8*	Study in state X; no controls for SES
Travis et al. (1990)	Prospective study; medical records	1.5	Women enrolled in prenatal care clinics in NYC; low SES only

RR: Relative risk.

SES: Socioeconomic status.

* $p < 0.05$.

Figure 12.4. Slide with tabular presentation of literature review

NHANES III Data

- 1988–1994 National Health and Nutrition Examination Survey
  - Nationally representative sample of United States
    - Oversample of Mexican Americans
  - Cross-sectional
  - Population-based
- $N = 9,813$
  - Response rate = 93%

Racial composition of sample

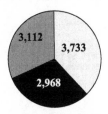

- ☐ Non-Hispanic white
- ■ Non-Hispanic black
- ▨ Mexican American

Figure 12.5. Slide describing data source using text and a pie chart

---

Variables

- Birth weight
  - Reported by mother at time of survey
  - Asked whether "low birth weight"
    - "Low" not defined on questionnaire
  - Also asked in pounds or grams
    - Classified LBW if < 2,500 grams
    - Measure used in our analyses
- Maternal smoking
  - Did she smoke cigarettes while pregnant?

- Socioeconomic status
  - Mother's education (years)
    - % < high school education
  - Mother's age at child's birth (years)
    - % teen mother
  - Family income-to-poverty ratio (IPR)
    - Family income in $ compared against poverty level for family of same size and age composition
    - % poor = IPR < 1.0

Figure 12.6. Slide describing major variables in the analysis

variables such as income or age that could be defined or classified in any of several ways, mention units or categories and explain how your measures of those concepts are defined.

Consider including a schematic diagram to illustrate how your variables are hypothesized to relate to one another (figure 12.7)—showing mediating or confounding relations, for example.

RESULTS

Portray univariate distributions for your main variables using charts such as the pie chart in figure 12.5 (which shows the distribution of the key predictor), or a histogram to show distribution of an ordinal variable. Then progress to slides portraying the bivariate association between your key predictor and outcome variables (figure 12.8) and other important bivariate or three-way associations. To facilitate a GEE summary, the relationships between race/ethnicity and each of the three socioeconomic variables are presented in one clustered bar

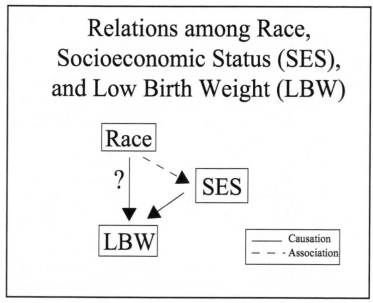

Figure 12.7. Slide with schematic diagram of hypothesized relationship among major variables in the analysis

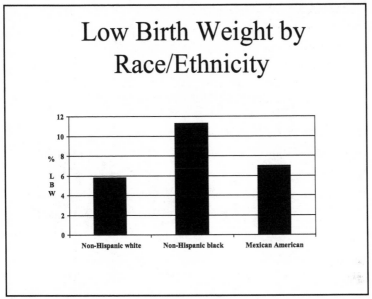

Low Birth Weight by Race/Ethnicity

%
L B W
Non-Hispanic white   Non-Hispanic black   Mexican American

Figure 12.8. Chart slide of main bivariate association
Source: US DHHS 1997

chart (figure 12.9) rather than as three different bar charts each on a separate slide.

Conclusions

Summarize your conclusions in one or more text slides such as figure 12.10, relating the findings back to your original research question or hypotheses, pointing out new questions that arise from your results, and discussing the policy and research implications of those findings. See "Data and Methods in the Discussion Section" in chapter 10 for more ideas.

General Guidelines for Slides
"KISS"

"Keep it simple, stupid," to reiterate one of the principles from chapter 2. Design each slide to concentrate on one or two major points, with title and content to match. Doing so divides your material into small, readily digestible chunks that are easier to organize into a logical,

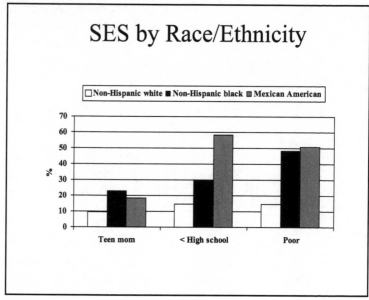

Figure 12.9. Chart slide to support a "GEE" pattern summary
Source: US DHHS 1997

Conclusions

- Much of racial/ethnic difference in LBW due to
  SES:
  - Infants of color more likely to be low SES.
  - Low SES infants more likely to be LBW.
  - When SES taken into account, racial differences in
    LBW differences narrow.

- Mexican American infants do better than expected
  despite low SES.
  - "Epidemiological paradox"

Figure 12.10. Text slide summarizing major study conclusions

straightforward sequence. Simple, uncluttered slides have another advantage: each can be covered in a minute or two—a much better way to maintain your audience's attention than showing the same crowded slide for several minutes while you slog through each of its contents.

HOW MANY SLIDES?
Figure on an average of one slide per minute, then err on the low side to avoid rushing and to permit time for questions or discussion. Although a simple text slide can often be covered in 30 seconds, those showing complex patterns or several specific facts may require several minutes apiece. If you are drafting a talk from a written document, start by creating one slide for each major paragraph or topic to be discussed. For short talks, be parsimonious in selecting what material to cover: a five-minute talk obviously cannot accommodate one slide for every paragraph and table in a 20-page document. Determine which parts of the paper are essential for introducing and answering the key points you have identified for that audience and time limit, then design slides accordingly.

Slide Formats
Like written documents, slides can include text, tables, graphs, diagrams, maps, and other types of graphical images. To enhance the visual appeal of your slides and introduce texture into your talk, vary the design of your slides to include a combination of these elements.

SLIDE TITLES
Good titles guide listeners through your talk, introducing the specific purpose of each slide and orienting listeners to the different sections of the talk. To outline a new speech or revise an existing talk for a new audience, write the titles for each of your slides before you fill in the body of the slide. Give each slide a short, specific title to identify the objective or content of that particular slide. General titles such as "Introduction" or "Results" tend to be ignored if they are repeated for several consecutive slides. The title features prominently on each slide—at the top in large type. Take advantage of that size and position: write informative titles! For instance, although the slides shown in figures 12.1 through 12.3 all comprise parts of the introduction,

their titles clearly identify which facet of the introductory material is covered in the respective slides.

Some speakers like to title each slide with a concluding point or rhetorical question related to the slide contents. For example, the title to figure 12.2 could be replaced with "LBW Stable over Past Two Decades" or "Has LBW Declined over Time?" Alternatively, put a title such as "Incidence of LBW" on the slide, then paraphrase it into a concluding point or rhetorical question as you introduce the slide.

TEXT SLIDES

Text slides can be used throughout a presentation, as an outline (figure 12.3), in the introduction (figure 12.1), in the data and methods section (figure 12.6), and in the discussion and conclusions (figure 12.10). Text slides also work well to summarize a few key points from previous studies, state hypotheses, list major results, or provide an executive summary. As you design each text slide, put vital numbers in a prominent position in large type, and make sure to report and explain them before they are used in any calculations or discussion. A NASA presentation about possible explanations of damage to the shuttle *Columbia*'s wing during its fatal flight placed critical numeric information in a footnote on the last slide where it was easily overlooked, making it hard to follow the logic of the investigation or understand its conclusions (Schwartz 2003).

Resist the urge to cut and paste sentences from a written document or speaker's notes into your slides. Instead, simplify your paragraphs and sentences into bulleted text phrases, aiming for no more than six bullets per slide and no more than six to ten words or numbers per bullet (Fink 1995; Briscoe 1996; Kosslyn 2007). These guidelines force you to plan simple, focused slides, and enhance readability by permitting large type and ample white space.

Bullets

Create a separate bullet for each concept, definition, or fact. Revise sentences into bulleted format in the following ways.

- Include only the essential words from each sentence—nouns, verbs, adjectives, and adverbs.

- Look for commas or the words "and" or "or" to identify clauses or elements of a list, each of which can become its own bullet item.
- Substitute common mathematical symbols such as <, >, =, #, or % for their equivalent phrases.
- Use arrows to convey directionality and causation.
- Eliminate most other words from the slide.
- Cast all bulleted points in the same syntax. If one is a sentence, make all sentences. Make all bullet points either active or passive, and use a consistent tense throughout. It's much easier to take in and remember points conveyed in a consistent, predictable form.

After you have drafted a bulleted version of a sentence or paragraph, review it to see whether more words can be eliminated without loss of meaning, or if additional words are needed to maintain clarity. Full sentences can be used in the accompanying speaker's notes.

Indenting

Use indenting to organize the material on a slide, presenting supporting facts or clusters of related information under one heading. In figure 12.6, socioeconomic status (SES) is one of several conceptual blocks of variables in the analysis. Indented below the bullet "socioeconomic status" is a list of the different continuous measures of SES, with one variable per bullet. Indented yet again beneath each of the SES measures is the categorical version used in this study to indicate low SES.

Observe how these principles improve the introductory slide shown in figure 12.11.

Poor: Figure 12.11.
   The slide includes the full text sentences from the introductory paragraph of the paper upon which the talk is based. Although each sentence is given its own bullet, the sentences crowd the slide and encourage viewers to read rather than listen. If it were in 12-point type, pasted in from the paper, it would be even worse! (Figure 12.11 is in 28-point type.) The title of the

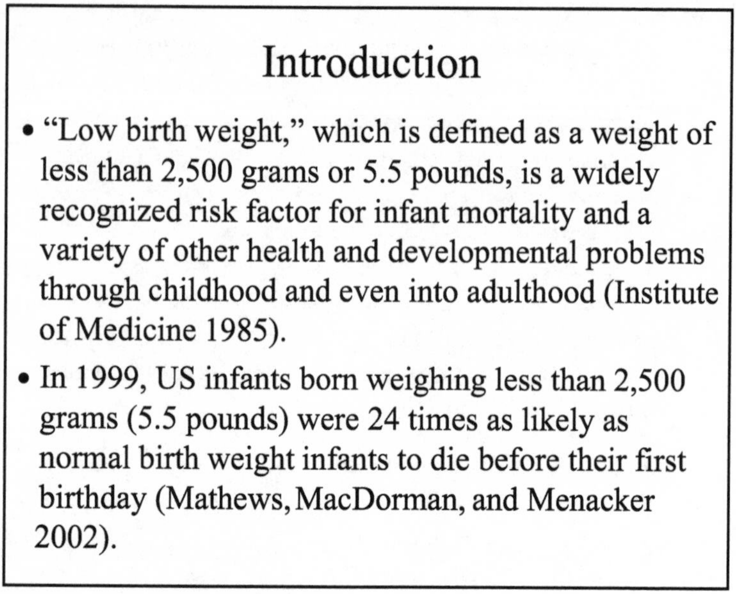

> # Introduction
>
> - "Low birth weight," which is defined as a weight of less than 2,500 grams or 5.5 pounds, is a widely recognized risk factor for infant mortality and a variety of other health and developmental problems through childhood and even into adulthood (Institute of Medicine 1985).
> - In 1999, US infants born weighing less than 2,500 grams (5.5 pounds) were 24 times as likely as normal birth weight infants to die before their first birthday (Mathews, MacDorman, and Menacker 2002).

Figure 12.11. Example of a poor introductory slide

slide describes its position in the talk but does not identify the contents or issues addressed.

Better: Figures 12.1 and 12.2.

This version includes the essential information from the poor version but is more succinct and better organized. The titles clue listeners into the specific topic and purpose of each slide. Clauses are broken into separate lines, with supporting information indented.

For an academic audience, mention key citations in the bullets or as footnotes. For lay audiences, omit citations except for public figures or widely recognized authorities (e.g., the Centers for Disease Control and Prevention).

Another example of text slide design, this time from the data and methods:

Poor: Figure 12.12.

Again, paragraphs are pasted directly from the paper onto a slide, resulting in an overcrowded slide that is difficult to read.

Better: Figure 12.5.

The slide title names the data source with an acronym that is spelled out later on that slide. The information from figure 12.12 is broken up into manageable pieces. Racial composition of the sample is presented in a pie chart, and the W's and other background information on the data source for the analysis are organized using bullets and indenting.

DIAGRAMS, MAPS, AND GRAPHIC IMAGES

In many cases a picture is worth a thousand words—a particularly valuable saving in a timed speech. Schematic diagrams can help viewers understand hypothesized relationships among variables (e.g., figure 12.7), using different types of arrows to illustrate association and causation. Timelines can portray the sequence of events under study or illustrate the number and timing of data collection points in a longitudinal study (e.g., figure 10.1). If your topic has an important geographic component, include one or more maps to present statistics such as population density or pollution levels for each area, or to

---

Data

- The data were taken from the 1988–1994 National Health and Nutrition Examination Survey (NHANES III), which is a cross-sectional, population-based, nationally representative sample survey of the United States. To allow for an adequate number of Mexican Americans to study separately, that group was oversampled in the NHANES III.

- Our study sample included 9,813 infants, including 3,733 non-Hispanic white infants, 2,968 non-Hispanic black infants, and 3,112 Mexican American infants.

Figure 12.12. Example of a poor data slide

show where the sites you discuss are located relative to hospitals, rail lines, or other features that pertain to your research question. Photographs of people or places can provide a richness difficult to capture in words.

ADAPTING CHARTS AND TABLES FOR SLIDES

Use slides with tables, charts, or other graphical material in both brief, general speeches and longer, in-depth presentations. Simple tables of numeric results work well for both scientific and applied audiences. For a scientific talk, a table that organizes and compares key literature on your topic can be very effective (e.g., figure 12.4).

Rather than use tables or charts that were designed for a written document, adapt them to suit a slide format. If your table or chart includes information on more than a few variables, it is impossible to discuss all the patterns simultaneously, so don't ask your viewers to ignore most of a large table or complex chart while you describe one portion. Instead, create two or more slides with simpler tables or charts, each of which includes only the information needed for one comparison. Although many publishers set limits on the number of charts or tables in a published document, such restrictions don't affect speeches, so take advantage of that flexibility by creating chart and table slides that concentrate on one or two straightforward relationships apiece.

First, identify the patterns you plan to discuss from a given table or chart, then design simplified versions that focus on one or two major points (or one GEE) apiece. Replace standard errors or test statistics with symbols or formatting to identify statistically significant results.

Poor: Figure 12.13.

The type size for the table, which was copied and pasted directly from the accompanying paper, is far too small for a slide. Even if you circle or highlight the numbers to which you refer, it is difficult for viewers to find (let alone read) those numbers and their associated labels. To describe the many patterns on this slide, you would have to ask viewers to wade through a lot of

Results

Birthweight, socioeconomic characteristics, and smoking behavior by race/ethnicity, US, 1988-1994

	Non-Hispanic white (n=3733)	Non-Hispanic black (N=2968)	Mexican American (N=3112)
Birth weight			
Mean (grams)	3,426.8	3,181.3	3,357.3
% low birth weight	5.8%	11.3%	7.0%
SOCIOECONOMIC CHARACTERISTICS			
Mother's age			
Mean (years)	26.6	24.2	24.9
% teen mother	9.4%	22.9%	18.4%
Mother's educational attainment			
Mean (years)	12.3	11.9	9.1
% < high school	14.7%	30.1%	58.4%
% exact high school	24.9%	41.7%	24.5%
Income/poverty ratio			
Mean	2.60	1.39	1.34
% poor	14.7%	48.5%	50.7%
HEALTH BEHAVIOR			
Mother smoked while preg.			
% smokers	26.8%	22.9%	10.1%

Figure 12.13. Example of a poor results slide
Source: US DHHS 1997

microscopic numbers and labels to find the three numbers that pertain to each comparison. E.g., "The second row of numbers shows the percentage of low birth weight births in each racial ethnic group..." [Description of that pattern.] [Then] "The fourth row of numbers shows the percentage of mothers in each racial/ ethnic group who gave birth as teenagers," etc.

Better: Figures 12.8, 12.9, and 12.14 .

The results from the table have been transformed into three separate slides, each of which presents data for one aspect of the story. Although this approach results in more slides, it takes no longer to describe because the amount of material is unchanged. It may even save time, because less guidance is needed to find the pertinent numbers for each comparison. The title of each slide names the variables or relationships in question. Speaker's notes would introduce each slide by identifying the role of the variables before describing the pattern and the findings on the topic at hand.

Maternal Smoking by Race/Ethnicity

	Smoked cigarettes (%)
Non-Hispanic white	26.8
Non-Hispanic black	22.9
Mexican American	10.1

Figure 12.14. Example of a better results slide: Simplified table created from larger table
Source: US DHHS 1997

- Figure 12.8 uses a simple bar chart to show how the outcome—low birth weight—varies by race and ethnicity.
- Figure 12.9 shows how each of the three socioeconomic variables relate to race/ethnicity. To facilitate a "generalization, example, exceptions" (GEE) summary, those patterns are presented in one clustered bar chart rather than as three different bar charts each on a separate slide.
- To emphasize that smoking is a behavior, not a socioeconomic variable, statistics on racial patterns in maternal smoking are shown on a separate slide (figure 12.14).

Although figure 12.14 presents the smoking pattern in tabular form, if I were presenting the material in an actual talk, I would replace the table with a bar chart. Why? Once you have introduced your audience to a certain format—in this case a bar chart—save time and minimize confusion by reusing that format throughout the talk whenever suitable.

If your charts or tables are fairly clear-cut (e.g., a 2-by-2 table, or a pie, line, or simple bar chart), consider a "chartbook" layout: a table, chart, or other image occupies one side of the slide, with bulleted text annotations on the other side (e.g., figure 12.2 or figure 12.5). Put more complicated tables or charts alone on a slide, then describe the pattern in your speaker's notes or make an additional slide with a short written summary.

DESIGN CONSIDERATIONS

Substance over Style

Resist the temptation to let the features available in presentation software packages carry your show. Fancy, multicolored background designs, animated text, or sound effects might impress your audience for a moment or two, but if they distract from your story line or substitute for correct, clearly presented material, they will do more harm than good. Whatever time you put into creating a dog-and-pony show is taken away from selecting and organizing the information and writing a clear narrative.

Focus on the substance, not the style, of the slides. First, get the content and organization right, just as you would for a written description of the same material. After you have practiced and revised your talk (see below), consider adding a bit of color or animation *only if they enhance your presentation.*

Color

That said, judicious use of color can enhance communication appreciably, giving you another tool for conveying information. For instance, use red type to identify all of the statistically significant findings in tables or text slides, leaving nonsignificant effects in a neutral shade. Once you have explained that color convention, your viewers will quickly be able to ascertain results of all your statistical tests without further explanation. Use a consistent color scheme for all charts within a talk. If the Northeast is represented in green in a pie chart illustrating sample composition, for example, use green for the Northeast in all subsequent charts (whether pie, bar, or line charts) that compare patterns across regions.

A caution about creating handouts from color slides: some color combinations and lighter colors do not reproduce well in grayscale—the typical color scheme for photocopied handouts. To make sure the handouts convey the same information as the projected slides, follow the guidelines in chapter 7 about using color in charts, then review them in black and white on-screen or in print before making copies. Or if your budget and equipment permit, make color handouts.

Type Size

Use a large type size on all slides—at least 18-point type—and avoid fussy calligraphic fonts. For your slides to be of value, they must be readable in the back row. If you aren't sure about the size of the room in which you'll be speaking, err on the generous side when you select your type size (see Zelazny 2001 for specific guidelines). If material you had planned for a single slide will fit only if you use small type, divide that material across several slides until the contents can be displayed with readable type.

Ditto for words used to label charts and tables: Think of all the talks you've attended where the speaker puts up a table or figure and says "you probably can't see this, but . . ." Why show an exhibit that your audience can't read? Instead, manually increase the type size for tables or charts you have pasted in from other documents or programs, again aiming for a minimum of 18-point type for all table elements, chart titles, axis labels, and legends. If those elements aren't editable, use a "zoom" feature to enlarge those components as you describe them during your talk.

Even with large type, slides can be difficult to read from the rear of a large auditorium. For such situations, consider printing handouts of your slides; some presentation software can print several slides per page with space for listeners to take notes.

Symbols and Annotations

As you adapt charts or tables for your slides, omit any features such as symbols, reference lines, or other annotations that you don't explain or refer to during your speech. Unless you mention them, they distract your viewers and clutter the slide. Conversely, you may want to add symbols to charts and tables as you modify them for use on slides. For example, your audience won't have time to digest detailed standard errors or test statistics during your talk, so replace them with symbols for $p < 0.05$ or $p < 0.01$ to save space and reduce the amount of data on the slide.[1] Include footnotes or legends to explain the symbols.

HANDOUTS

A question that often arises is whether to hand out statistical tables like those from the printed paper. (By now, you've probably figured out not to give such tables to lay audiences.) Unless I am presenting at a long seminar where active audience discussion of detailed results is expected, my preference is to distribute such tables after the presentation. This approach gives readers the full set of numbers to peruse at their leisure without getting distracted during the talk. If you want to give handouts for viewers to follow along with a shorter speech, include copies of the most critical slides. Coordinate the handouts with your slides so you and your viewers are looking at the same set of materials as you speak.

Poor (for a short speech): A hardcopy of a detailed statistical table
(e.g., table 11.1), unaccompanied by any slides.
Although you could try to describe each number's position
(e.g., "in the second row from the bottom, third column labeled
'Mexican American,' do you see the value 50.7?"), without a slide,
you don't have anything to point at to guide your viewers.
Poor (version 2): A handout of a detailed table (e.g., table 11.1)
accompanied by simplified slides such as figures 12.8
and 12.9
If the hardcopy and the slides don't look alike, you will squander
even more time guiding readers to the portion of their handout
that corresponds to the material on each slide. If you use a
handout, make it match your slides, or distribute the detailed
tables at the end for later inspection.
Better: Figures 12.8 and 12.9 shown on the screen and reproduced
in a handout.
By dividing the analytic points across several slides, you can
describe each issue in turn, referring to numbers that you point
out on a simple table or chart.

Having created slides that present the essential textual and graphical elements of your talk, write speaker's notes to fill in the details and transitions among slides. Although you can draw heavily on the content and organization of a paper or book when formulating these notes, avoid recycling large blocks of text in your speech. Rarely will you be able to read an entire paper in the time available. Even if time permits, reading a document out loud is a poor substitute for a speech.

Speaker's Notes to Suit Your Style

The notes can be adapted to suit your speaking style and level of experience. If you are a novice, are uncomfortable inventing sentences in front of an audience, or have a tendency to be long-winded, you might do best with a full script. The wording for such notes can be pirated largely from the corresponding written paper or article, cutting some of the detail (such as citations) and rephrasing into the first person. For figure 12.5, a script might read:

> "We used data from the 1988–1994 National Health and Nutrition Examination Survey, also known as NHANES III, which is a cross-sectional, population-based, nationally representative sample of the United States. To allow for an adequate number of Mexican Americans to study separately, that group was oversampled in the NHANES III. We excluded infants of racial and ethnic groups not shown on this slide because we did not want to group them with any of these three groups and there were not enough of them to analyze as a distinct group. As shown in the pie chart, our study sample comprised nearly ten thousand infants, approximately equally distributed among the three racial/ethnic groups studied."

If you are at ease speaking extemporaneously and are able to keep yourself on schedule, you might need only a list of additional points to make or items to underscore. For the same slide, such notes might read:

"To allow for an adequate number of Mexican Americans to study separately, that group was oversampled in the NHANES III."
"As shown in the pie chart, our study sample comprised approximately equal numbers of the three racial/ethnic groups studied."

Before reading those notes, restate the information in the title and bullets into two or three complete sentences. Using selected reminders takes more practice than working from notes that comprise the full speech because you must remember where each typed note fits within the overall description of each slide. Key your notes to your slides to coordinate the spoken and visual components of your speech. Some presentation software programs allow you to type speaker's notes for each individual slide. Do yourself a favor and print your speaker's notes in large type so you won't have to squint to read them as you deliver your speech. If the notes won't fit with the larger type size, you are probably trying to say too much about the slide. Instead, decide whether to divide the material across several slides or sacrifice some of the less-vital content from your talk. If you write your notes longhand or in a word processor, write the number of the slide, table, or chart in the margin next to the associated text to remind yourself when to change slides.

Explaining a Chart "Live"
Tables, charts, maps, and other diagrams offer real advantages for presenting numeric patterns. Unfortunately, many speakers devote far too little time to describing such slides. They put up the slide, state "as you can see . . ." and then describe the pattern in a few seconds before moving on to the next slide. As the slide disappears, many listeners are still trying to locate the numbers or pattern in question and have not had time to digest the meaning of the statistics. This disease plagues rookie and veteran speakers alike: Beginners may not want to spend very long on a chart out of fear that they will run out of time (or because they just want to get their talks over with). Experts forget that not everyone is conversant with their chart or table layouts or may be too uppity to explain such rudiments.

Although it may appear to save time, failing to orient your listeners to your charts or diagrams reduces the effectiveness of your talk. If you designed the chart and wrote the accompanying talk, you know it well enough to home in quickly on the exact number or table cell or trend line you wish to discuss. Give your audience the same advantage by showing them where to find your numbers and what questions they address before you report and interpret patterns.

Follow these three steps to explain a chart, table, or other diagram as you give a speech.

INTRODUCE THE TOPIC

First, state the topic or purpose of the table or chart, just as you do in the introductory sentence of a written paragraph. Rather than read the title from the slide, paraphrase it into a full sentence or rephrase it as a rhetorical question. For figure 12.9:

> "This slide examines racial and ethnic patterns in each of three indicators of low socioeconomic status. In other words, 'Does socioeconomic status vary by race?'"

EXPLAIN THE LAYOUT

Second, explain the layout of the table or chart. Don't discuss any numbers, patterns, or contrasts yet. Just give your audience a chance to digest what is where. For a table, name what is in the columns and rows. For a chart, identify the concepts and units on the different axes and in the legend, mentioning the color or shading of bars or line styles that correspond to each major group you will discuss. For maps or other diagrams, point out the location of different features and explain the meaning of legend items or other elements such as arrows, symbols, or scales.

Use a "Vanna White"[2] (or TV meteorologist) approach as you explain the layout, literally pointing out the applicable portion of the table or chart as you mention it. Point with a laser pointer, pen, or finger, or use an animated circle or arrow to highlight the pertinent feature—it doesn't matter. The important thing is to lead your viewers' eyes across the key elements of the slide before reporting or interpreting the information found there. At first this may seem silly or awkward, but most audiences follow and retain the subsequent description much more easily than if you omit the guided tour. See online materials for a demonstration.

Below, I use bracketed comments to describe the Vanna White motions that accompany the surrounding script; they are there to guide you, not to be spoken as part of the presentation. For figure 12.9:

"Across the bottom [wave horizontally at the x axis], there is one cluster for each of the three socioeconomic characteristics—teen motherhood, incomplete high school, and poverty [point quickly at each label in turn]. Each racial/ethnic group [point to the legend] is displayed with a different color bar, and the height of a bar [gesture vertically along the y axis] shows the percentage of that racial or ethnic group with the associated characteristic."

For figure 12.15:

"The income-to-poverty ratio is shown on the x axis, ranging from zero to four times the poverty level [wave across x axis]. The y axis [point] shows the increment in birth weight in grams compared to infants in families with an income-to-poverty ratio of 0, taking into account the other variables listed at the bottom of the slide [point]."

For figure 7.11:

"In figure 7.11, there is one bar for each racial/ethnic group (color) [point to legend] within each educational attainment group (cluster) [point to clusters on x axis]. The height of each bar is the mean birth weight in grams for infants in that group [wave vertically along y axis]. The 95% confidence interval around each mean birth weight estimate is shown by the vertical error bar that brackets the top of each bar [point to top and bottom of one confidence interval]."

In the next step, you will give a specific example and introduce the bar colors for each subgroup. For lay audiences, "x axis" and "y axis" may be fuzzily recalled jargon. Instead, use phrases like "across the bottom" or "on the vertical axis," respectively.

If you are explaining a chart with more than three or four nomi-

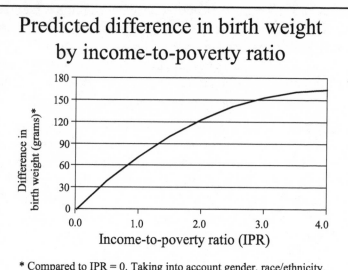

Figure 12.15. Chart slide to show non-linear pattern
Source: US DHHS 1997

nal variables or categories, mention the organizing principle you have used rather than simply naming each of the categories. As always, coordinate the narrative with the layout of the chart (chapter 7).

"In figure 7.6, the different AIDS transmission topics are shown on the horizontal axis [point] grouped into 'likely' modes of disease transmission on the left [wave at that group of clusters] and 'unlikely' modes on the right [gesture]. Within those groupings, the topics are arranged in descending order of average score [wave along the tops of the bars within one group of clusters]."

DESCRIBE THE PATTERNS
Finally, having introduced your audience to the purpose and layout of the table or chart, proceed to describe the patterns it embodies. Use the GEE approach, starting with a general descriptive sentence followed by specific numeric examples and exceptions (where pertinent). Again, gesture to show comparisons and point to identify

specific values, naming the associated colors or shading schemes for each group the first time you mention it, as shown in the following description of figure 12.9.

"Regardless of which dimension of socioeconomic status we examine, non-Hispanic black infants, illustrated with the black bar, and Mexican American infants—the gray bar [point at legend]—are far more likely than their non-Hispanic white counterparts, in white [point at legend element], to be born into low SES families. The black and gray bars are higher than the white bar in each of the three clusters. For example [gesture at the rightmost cluster], infants of color are more than three times as likely to be poor as their white counterparts [point to the respective bars as you mention them]."

For figure 12.15:

"As the income-to-poverty ratio (abbreviated IPR) increases, birth weight also increases, but at a decreasing rate. For example, predicted birth weight is about 51 grams heavier for infants born into families at twice the poverty level than for those at the poverty level [point to the IPR curve between IPR = 1 and 2]. In contrast, an increase in IPR from 3 to 4 is associated with a predicted increase in birth weight of only about 11 grams [gesture along curve between IPR = 3 and 4] as the curve levels off at higher income levels."

As you describe your charts, tables, or other graphics, point to and explain the purpose of features such as reference lines or regions, colors, symbols, or other annotations. For figure 7.11:

"The birth weight deficits for non-Hispanic black (black bar) [point] compared to non-Hispanic white infants (white bar) [point] within each education level were statistically significant at $p < 0.05$, as illustrated by the fact that their 95% confidence intervals do not overlap one another [within the <HS cluster, wave horizontally from the upper confidence limit for non-Hispanic black to the lower confidence limit for non-Hispanic white infants]."

For figure 7.14:

> "As you can see, recessions, which encompass the years in the gray-shaded bands in figure 7.14 [gesture at the range of shaded dates], coincided with a notable increase in poverty [wave along the line showing the trend in poverty]."
>
> "In table *yyy* (not shown), relationships that were statistically significant at $p < 0.05$ are shown in orange and are marked with asterisks [point to the footnote on the slide that defines the asterisk]. For example, the difference in average math scores between boys and girls was statistically significant [point to pertinent cells], but most other comparisons in the table were not."

Until you are confident that you can recall your Vanna White description, include it in your speaker's notes, either in full sentences or as circles and arrows on a hard copy of the chart, numbered to help you recall the order in which you plan to explain each feature.

PRACTICE, PRACTICE, PRACTICE

After you have drafted your slides and accompanying notes, practice your presentation, first alone and then with a test audience. If someone else wrote the speech and made the slides, all the more reason to review and practice. Rehearsal is particularly important for slides involving tables, charts, or other diagrams, which are usually more complex than simple text slides. Likewise for slides explaining methods, especially if you have not worked previously with those methods or explained them to a similar audience.

Time how long the entire talk takes, anticipating that you will become somewhat faster with practice (and adrenaline). If you will be using a Vanna White approach, rehearse speaking and gesturing at the associated chart until you are comfortable coordinating those two actions. Evaluate the order in which you've covered the material, making sure you define terms, acronyms, and symbols before you use them, and that your results are in a logical order with good transitions to convey where they fit in the overall story.

If you exceeded the allotted time by more than a minute or two, identify which sections were too long and assess what can be condensed or eliminated. Some sections will require more time than others, so you may have to omit detail or simplify explanations in other parts of your talk, taking into account what your audience knows (and needs to know). If you finished well under time, think about where additional material or explanation would be most useful. If you were under time but rushed your delivery, slow down.

Revise the coverage, level of detail, and order of material to reflect what you learned from your dry run. If you make substantial revisions, practice on your own again before you enlist a test audience. To assist yourself in pacing your talk, insert reminders in your speaker's notes to indicate where you should be at certain time points so you can speed up or slow down as necessary during your talk. As you are introduced at the talk, write down the actual start time and adjust these time points accordingly.

Dress Rehearsal

Once you have a draft of slides and notes that you are comfortable with, rehearse your talk in front of a colleague or friend who represents your audience well in terms of familiarity with your topic, data, and methods of analysis. If you differ substantially from your prospective listeners on those dimensions, it is difficult to "put yourself in their shoes" to identify potential points of confusion. A fresh set of eyes and ears will be more likely to notice such issues than someone who is jaded from working closely with the material while writing the paper or drafting the slides and talk.

Before you begin your dress rehearsal, ask your guinea pig audience to make notes on the following aspects of your talk:

- Were the objectives of your talk plainly identified?
- Were the purpose and interpretation of your numeric examples evident?
- Were your definitions of terms and concepts easy to grasp? Did you define terms before you used them?
- Did you use jargon that could be replaced by terms more familiar to this audience?
- Were your descriptions of tables or charts clear and not too rushed? Was it easy to see where your numeric examples

came from in those tables or charts? To follow the patterns you described?

- If you were over time, what material could be omitted or explained more briefly? If under time, where would more information or time be most beneficial?
- Was the amount of time for each section about right? If not, which sections need more or less emphasis?

Go over your reviewers' comments with them, then revise your talk and slides accordingly. Practice yet again if you make appreciable changes.

CHECKLIST FOR SPEAKING ABOUT NUMBERS

Before you plan your speech, consider your topic, methods, audience, and amount of time, pacing the talk for the average listener and allowing time for questions and discussion.

- Slide format and content: adapt material from your paper, following the same sequence of major topics.
    For a scientific audience, include an introduction, literature review, data and methods, results, discussion, and conclusions.
    For lay audiences, omit the literature review, and condense the data and methods.
    Write a simple, specific title for each slide.
    Replace full sentences with bullets.
    Simplify tables and charts to focus on one major question per slide.
    Create no more than one slide per minute, fewer if slides involve tables or charts.
- Speaker's notes: decide whether you need a full script or selected notes. In either case, follow these steps for each slide:
    Write an introductory sentence.
    Note aspects of the slide you want to emphasize.
    Include analogies, examples, or anecdotes you will use to flesh out the material.
    Write a Vanna White description of charts or tables.
    – Paraphrase the purpose of the slide.

- Explain the layout of the table (contents of rows and columns) or chart (axes, legend), with notes about which elements to point to for each sentence.
  - Describe the pattern, listing which illustrative numbers you will point to as you speak.

  Write a summary sentence.

  Insert a transition to the next slide.
- Other design considerations:

  Use at least 18-point type for titles and text including table contents and chart labels; slightly smaller for footnotes.

  Consider using color to emphasize selected points or terms, or to indicate statistical significance.
- Rehearsing your talk. First alone, then with a critic familiar with your intended audience, evaluate the following:

  Order and relative emphasis of topics

  Definitions of terms

  Level of detail

  Introductions and explanations of charts and tables

  Coordination of spoken and visual materials

  Time to complete the talk

1 2 3 4 5 6 7 8 9 10 11 12 **13**

WRITING FOR APPLIED AUDIENCES
ISSUE BRIEFS, CHARTBOOKS, POSTERS, AND GENERAL-INTEREST ARTICLES

Many potential audiences for writing about numbers are "applied audiences"—people who raise questions that can be answered using numeric facts but who are interested principally in the answers rather than the technical details of how they were obtained. Communicating about numbers to applied audiences is a routine task for consultants, policy analysts, grant writers, meteorologists, science writers, and journalists; academics and other researchers also must do so when addressing funding agencies or others from outside the research community.

Earlier in this book, I explained how to write papers or reports involving numeric information to scientific audiences, who can be expected to have some training in quantitative methods. In this chapter, I illustrate how to translate findings of those same analyses so they are comprehensible to a wide range of people, but without "dumbing down" the analyses themselves. I begin by discussing audience considerations, then cover how to adapt tables, charts, and text for applied audiences. I then discuss the contents, length, format, and style of several different formats that are commonly used for such audiences, including posters, chartbooks, issue briefs, and general-interest articles. I also illustrate how to write an executive summary, which can be used to accompany reports or chartbooks. These formats are also effective ways to present results of more advanced statistics including multivariate models; see Miller (2013a), chapter 20. For an indepth review of different formats and considerations for communicating to applied audiences, see Nelson et al. (2002).

WHY IS WRITING FOR APPLIED AUDIENCES DIFFERENT?

For most people, statistical analysis is not the end in itself, but rather a tool—a way to address questions about relationships among the concepts under study. For analyses to be genuinely useful and interesting to an applied audience, make your results accessible to people who understand the substantive context of your analysis, whether or not they "do statistics" themselves. A few common examples:

- Policy analysts must explain results of their analyses to experts in government or nonprofit agencies—professionals who are well-versed in the issues and their application, but few of whom are skilled statisticians. Their principal interest is in the findings and how to interpret and apply them, along with reassurance that you know how to use the statistics correctly.
- Economic consultants have to communicate results of their models to professionals in corporations, community development agencies, and other settings. Again, their clients bring many other pertinent types of knowledge to the table, but may not know much about statistics.
- Grant writers must explain their analyses to nonstatistical reviewers at charitable foundations as well as to both statistical and substantive reviewers at scientific research institutes such as the National Science Foundation. Nonstatistical reviewers will be interested in why the analytic technique you have chosen is needed to analyze the issues at hand, and how the results can be applied, rather than in details about the statistical methods.
- Science writers have to communicate findings to readers of the popular press.

For people trained in quantitative methods, nonstatistical audiences are often the most difficult to write for. Most of us who use these methods learned about them in courses that emphasized understanding statistical assumptions and calculations, some of which are pretty complex and arcane. Sensibly, the material in these courses is conveyed using a teaching style, demonstrating mastery by working with equations written in statistical notation and identifying the relevant numbers for formal hypothesis testing, with the expectation that the audience (students) will repeat each of these steps.

Readers with training in statistical methods can work from such shorthand and do much of the interpretation themselves, given the statistical output. However, nonstatisticians cannot be expected to extract the information they want from such raw materials, any more than most of us can create artisanal bread from a bag of ingredients. We expect bakers to make bread for us regardless of our expertise in (or ignorance of) their techniques. Most of us wouldn't be able to replicate their work even given a detailed recipe, and we don't want a lecture on how to bake every time we buy a loaf of bread. Likewise, applied audiences expect us to put our results in a format they can appreciate regardless of their proficiency (or lack thereof) with mathematical or statistical methods. Few of them could reproduce our analyses even given meticulous instructions, and they don't want a sermon on statistics every time they hear about the results. Just as we would be grateful for some interesting serving suggestions for an artisanal loaf of bread, applied audiences will welcome basic guidance about how to understand and apply results of your analyses.

There is a wide range of statistical training and interest among applied audiences.

- Some readers want to review the assumptions and methods behind the analysis, just as some people want to learn alongside an experienced baker which ingredients or techniques yield bread that suits their tastes. If you are conducting a study to forecast growth rates in a particular industry, explain to your client which assumptions and variables you plan to include so they can give you feedback about whether that approach is consistent with their understanding of the subject.
- Some readers want to hear a bit about the analytic approach, just as some people want to know the ingredients and techniques their baker uses so they can assess whether it follows the latest nutritional guidelines. If you are writing about determinants of student achievement for a group of

professional educators who have heard of multivariate models and have a general sense of why they are important, mention that such a model was estimated and which variables were included (Miller 2013a, chapter 20).

- Some readers want to know only what questions you addressed and what you concluded, just as some people simply want to enjoy their bread without worrying about its nutritional content or how it was made. If you are giving a five-minute speech to your neighborhood association, concentrate on the concepts and findings, not the statistical methods or variables.

Before you adapt your writing, assess the interests, statistical abilities, and objectives of your audience. If you will be presenting your findings to several different audiences, plan to create several different versions.

WRITING FOR APPLIED AUDIENCES

In general, writing for an applied audience involves greater prominence of the research question, reduced emphasis on technical details of data and methods, and translation of results to show how they apply to real-world issues of interest to that audience. Your main objective is to write a clear, well-organized narrative about your research question and answers. Explain in plain English what you did, why you did it that way, and what it means. For very short formats such as issue briefs or general-interest articles, some of these topics will be omitted; see pertinent sections below.

Giving an Overview of Data and Methods

Descriptions of data and methods are typically much shorter and less technical in documents written for applied than for scientific audiences. With the exception of posters for statistical audiences, translate all jargon and every statistical concept into colloquial language, and resist the urge to include equations or Greek symbols. For readers who are interested in the technical details, provide a citation to the scientific papers or reports on your analysis.

In the next few paragraphs, I provide guidelines and examples for length, content, and style of presenting essential information on data and methods to applied audiences.

SPECIFYING THE CONTEXT

To set the context of your study for an applied audience, incorporate the W's (who, what, when, where) and units as you write about the numbers rather than in a separate section.

REPORTING DATA SOURCES

For many applied audiences and short formats, data sources can be mentioned as you present the information. For example, "A report by the Federal Aviation Administration showed . . ." or "According to statistics from the World Bank . . ." In a chartbook or issue brief, provide a citation to that source. General interest articles usually do not include a reference list.

If more detail is needed to describe how a variable was collected, use everyday language rather than technical terminology.

> *Poor, for an Applied Audience:* "A doctor's report of asthma was based on (1) checking that diagnosis on a list of possible diagnoses, (2) listing 'asthma' on the open-ended section of the medical record, or (3) listing an IDC9 code of 493 on the open-ended section of the medical record (NCHS 1995)."
> *Lay readers might not know (and probably don't need to know) what an "open-ended question" or an "ICD9 code" is.*
> *Better, for an Applied Audience:* "The doctor's measure of asthma was based on whether a physician wrote 'asthma' on the medical record or checked it on a list of possible diagnoses."
> *This version conveys the same ideas in familiar language.*

DEFINING MEASURES AND VARIABLES

As you introduce the variables and measures used in your study, rephrase statistical terminology to focus on the underlying ideas, provide a definition, or show how the concepts in that measure apply to your particular topic. To report a type of statistic that is unfamiliar to your audience, embed the definition in your explanation:

Poor: "The sensitivity of the new screening test for diabetes is 0.90."
People who do not routinely study screening tests may not know
what sensitivity measures or how to interpret the value 0.90.
Better: "The new screening test had a sensitivity of 0.90, correctly
identifying 90% of diabetics."
This version clarifies both the metric and purpose of "sensitivity."

Presenting Results

As you present numeric results, follow the basic principles for reporting context, units, and categories, and for both reporting and interpreting patterns explained in chapters 2 and 9, and Miller (2006). Use
a combination of numeric contrasts and vocabulary to communicate
the shapes of patterns (Miller 2010).

UNIVARIATE PATTERNS

Report simple summary statistics on the main variables under study
such as mean or modal values of your main variables in prose, substituting "average" for "mean" and "most common" for "modal." For continuous variables, describe variation around the mean using phrases
like "spread" in place of "standard deviation." Charts can be extremely
effective in getting across the shape of a distribution, accompanied by
and phrases like "bell-shaped," "flat," or "smile-shaped." See the section on "Univariate Distributions" in chapter 9 for examples of how to
describe distributions of nominal and ordinal variables.

BIVARIATE ASSOCIATIONS

As you present results of bivariate associations such as time trends,
correlations, cross-tabulations, or differences in means, focus as always on both reporting and interpreting the patterns. Convey the direction and magnitude of differences in values using results of basic
arithmetic or statistical calculations explained using nontechnical
language.

"As expected, children who scored well on one test of academic
achievement also typically scored well on achievement tests
in other subjects. Correlations between achievement and

aptitude were generally lower than those among different dimensions of achievement."

This description summarizes the direction of association between the two broad concepts identified in the sentence, and replaces names of specific statistical measures with their conceptual equivalents.

See chapter 9 for additional examples of effective ways to present results of cross-tabulations or a difference in means; chapter 5 for suggested ways to phrase results of different types of quantitative comparisons including subtraction, division, and percentage change.

THREE-WAY ASSOCIATIONS: INTERACTIONS

Three-way associations are those involving three or more variables together, such as how a series of outcomes each relate to the same predictor (e.g., knowledge of AIDS topics by language, in figure 7.6), or how two predictors together relate to an outcome (e.g., how race and gender are jointly associated with labor force participation; figure 9.2).

One variant of such a pattern occurs when the size or shape of an association between one predictor variable and the outcome differs depending on the value of a second predictor. Statistically-oriented audiences often refer to such patterns as an "interaction," so for them, using that term to introduce the pattern makes sense.

For a Scientific Audience: "Race and gender interact in their association with labor force participation (figure 9.2)."

For an applied audience, you need not use the term "interaction" at all—a genuine advantage when writing for people who aren't familiar with the statistical meaning of that word. Write:

"As shown in figure 9.2, the relationship between race and labor force participation depends on gender."

Or

" As shown in figure 9.2, the relationship between race and labor force participation is different for men than for women"

Having alerted your readers to the fact that the pattern of association cannot be captured with a simple generalization, progress through the rest of the summary, using the "generalization, example, exception" (GEE) technique (chapters 2 and 9 and appendix A). This approach is easily understood by most audiences because it introduces the substantive patterns before illustrating them with numbers from the associated table or chart.

COMPLEX PATTERNS

Even results of complex patterns or multivariate models can be explained in ways that will not scare off nonstatistically-oriented readers. For instance, a *New York Times* article about a model to predict age-related changes in runners' marathon times wrote:

> "'I'm right now at the age where things are getting worse in a bigger way,' said Dr. Fair, using colloquial language to describe the increase in the second derivative on his chart."
> (Leonhardt 2003)
> *This excerpt phrases the statistical concept in everyday language, then ties that wording to the more technical language that statisticians would use.*

See Miller (2013a) for extensive guidance on presenting results of ordinary least squares, logistic regression, and other types of multivariate models to audiences that are not trained in those methods.

Statistical Significance

For applied audiences, keep the discussion of statistical significance simple, stressing the conclusions as they apply to your particular research question, not the computational process or logic. No matter how carefully you try to phrase it, a discussion of the purpose and interpretation of statistical tests may confuse readers who are not trained in statistics. Focus on issues related to your main research question and emphasize only statistically significant differences. Those that are not statistically significant should be described as such or omitted. Provide verbal description rather than numeric results of statistical tests:

- "The new math curriculum raised average test scores by five points in high-income schools."
- "Average test scores in low-income schools were not affected by the new curriculum."
- "Smaller classes performed slightly better than large classes."
- "There was no difference between racial groups in math scores within schools of similar income level."

If you anticipate that your audience needs to know results of inferential statistical tests, use those tests to screen which results you report and how you discuss findings rather than reporting standard errors or test statistics. Consider these examples of how to paraphrase the concepts behind the statistics into everyday language.

Poor: "In 2010, the mean score on the mathematics test for fourth graders in School A was 62.7% correct. The mean score on the same test in School B was 72.0% correct. The standard error of the difference in means was 2.9. Because the difference in means (9.3 percentage points) is more than twice the standard error of the difference, we conclude that the difference cannot be attributed to random variation in scores at the two schools."
This description puts too much emphasis on the logic behind the statistical test. Skip the statistics lesson and just report whether the difference between the two schools' test scores is statistically significant.

Better: "On a mathematics test given to fourth graders in 2010, students in School A achieved a lower average score than students in School B (62.7% and 72.0% correct, respectively). The chances of observing a difference this big in our study if there were no real difference between groups was less than one in a thousand."
This description reports the two scores and suggests that they represented different levels of achievement. Statistical significance is worded without reference to technical concepts such as p-values or test statistics.

If, contrary to previous evidence, the difference is not statistically significant, write:

"In 2013, Schools A and B achieved similar average math scores on a standardized mathematics test given to fourth graders (71.7% and 72.0% correct, respectively). These results run counter to findings from 2010, which showed appreciably weaker performance in School A than in School B. The difference in the schools' 2013 scores could easily have occurred by chance alone."

This version explains that in 2013 the two schools' scores were very close, and that the recent pattern differs from what was previously observed. Statistical significance is implied by the phrase "appreciably weaker scores" in the earlier study, and lack of statistical significance in the more recent study by the phrase "chance alone."

Recall the power of a metaphor or analogy for conveying an abstract statistical concept such as a 95% confidence interval by relating that concept to a concrete, familiar example.

Poor for a Lay Audience: "A confidence interval gives an estimated range of values which is likely to include an unknown population parameter, the estimated range being calculated from a given set of sample data. The level C of a confidence interval gives the probability that the interval produced by the method employed includes the true value of the parameter θ. For instance, a 95% confidence interval (C = 95%) is calculated as the point estimate of $\theta \pm (1.96 \times$ standard error of the estimate). If repeated random samples were taken from the population and the 95% confidence interval was computed for each sample, 95% of the intervals would contain the population mean."

Don't repel nonstatisticians with unfamiliar with terms like "parameter," "level C," or "standard error." Mathematical symbols and equations are also more likely to frighten off many lay readers, who might then believe they won't be able to follow your explanation.

Better: "To illustrate the idea behind a confidence interval, imagine someone who is going to throw a dart at the board and try to hit the bull's-eye. This is not a standard dart, however: in place of the point on the end is a suction cup. I succeed in hitting the bull's-eye if the suction cup covers it. Hence I have a margin of error which is the radius of the suction cup. With practice you can get good at this, perhaps good enough that 95% of the time you capture the bull's-eye under the suction cup (Weston undated)." The area encompassed by the suction cup dart is the 95% confidence region around the point estimate of the dart tip.

This version gets across the idea of a range capturing the true numeric value using an effective visual metaphor, and without unneeded technical information.

ADAPTING TABLES AND CHARTS

Simplify tables and charts from versions designed for academic audiences, replacing standard errors or test statistics with *p*-values, symbols, or boldface to denote statistically significant findings. If your analysis involves unfamiliar variables or complex mathematical transformations, paraphrase them or replace with a more familiar version. For instance, many lay readers will be more familiar with income in dollars than the income-to-poverty ratio (IPR), so in figure 13.1 the IPR is converted into the equivalent income for a family of two adults and two children based on the 1999 Federal Poverty Level, with a footnote to explain that translation. Readers who aren't conversant with the definition or value of the poverty level will find it easier to grasp the shape of the association from this figure than from the more technical version shown in figure 12.15.

Causality

As noted in chapter 3, for analyses intended to inform programs, policies, or other interventions, it is important to convey whether an observed association is causal, and to describe possible causal mechanisms. For example, readers need to know whether adopting a new math curriculum is the reason for improved math scores in those schools, or whether alternative explanations such as bias, confounding, or reverse causation might explain that pattern. Consider these two versions of presenting results of a hypothetical study to lay readers:

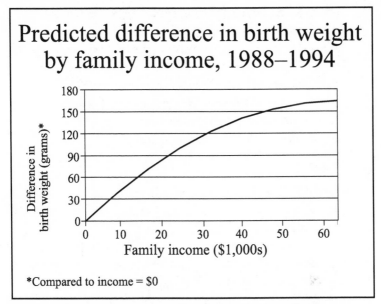

Predicted difference in birth weight by family income, 1988–1994

Difference in birth weight (grams)*

Family income ($1,000s)

*Compared to income = $0

Figure 13.1. Line chart to convey a non-linear pattern to a lay audience
Note: Taking into account gender, race/ethnicity, mother's age, educational attainment, and smoking status. For illustrative purposes, this figure uses the 1999 poverty threshold of $16,895 for a family of two adults and two children. Data are from US DHHS 1997.

Poor: "In 2013, School Q, which had adopted the new math curriculum two years earlier, averaged 10 percentiles higher on a standardized math test than School R, which continued to use an older curriculum."

By omitting any reference to the way the study was conducted and how that might affect interpretation of the data, this explanation implies that the new curriculum is the cause of Q's better performance.

Better: "In 2013, School Q, which had adopted the new math curriculum two years earlier, averaged 10 percentiles higher on a standardized math test than School R, which continued to use an older curriculum. However, School Q is in a higher income district that could afford the new curriculum and has smaller average class sizes and more experienced teachers than School R. Consequently, School Q's better performance cannot be attributed exclusively to the new curriculum."

By mentioning alternative explanations for the observed cross-sectional differences in math performance, this version explains that the evidence for beneficial effects of the new curriculum is relatively weak. A discussion of other study designs that could better assess causality would further strengthen this explanation.

Substantive Significance

As when writing for scientific audiences, differentiate between statistical significance and substantive importance by devoting attention to both aspects of your results. To convey the substantive meaning of your findings, place them in perspective by providing evidence about how they relate to some real-world outcome such as costs or benefits, for the status quo and other alternatives.

> *Poor:* "The association between race and birth weight was highly substantively significant."
>
> *Most readers won't know what "substantively significant" means. This version also fails to interpret the direction and size of the association, and doesn't provide insight into whether the change is big enough to matter.*
>
> *Better:* "What are the health and financial repercussions of the observed racial difference in birth weight from the NHANES III study? If the incidence of low birth weight among black infants could have been reduced to the level among white infants, nearly 40,000 of the low birth weight black infants born in 2000 would instead have been born at normal birth weight. That reduction in low birth weight would have cut the black infant mortality rate by more than one-third, assuming the infant mortality rate for normal birth weight infants (Mathews, MacDorman, and Menacker 2002). In addition, an estimated $3.4 million in medical and educational expenses would be saved from that birth cohort alone, based on Lewit and colleagues' estimates of the cost of low birth weight (1995)."

A rhetorical question and accompanying answer are an effective way to raise the issue of substantive significance, particularly in spoken formats. By pointing out that the elevated risk of low birth weight among black infants adds substantially to infant mortality, medical and educational costs, this version helps readers understand immediately why it is a pressing issue.

See "Substantive Significance" in chapter 3 for additional explication and examples.

GENERAL DESIGN CONSIDERATIONS

A few general guidelines for organization, length, and design apply to most formats for applied audiences. In later sections, I discuss how these considerations pertain to each of the different formats.

Organizing the Material

The W's are an effective way to organize the elements of reports, posters, or chartbooks.

- In the introductory section, describe what you are studying and why it is important.
- In the data and methods section, list when, where, who, and how the data were collected, how many cases were involved, and how the data were analyzed.
- In the body of the work, explain what you found, using tables or charts accompanied by short prose descriptions or text bullets.
- In the conclusion, explain how your findings can be applied to real-world issues or future research.

Briefs and general-interest articles are rarely divided into these formal parts; see sections below for information on how to organize them.

Text Length and Style

The amount and style of text varies from concise, bulleted phrases in a chartbook, to short paragraphs and bullets in a research poster or issue brief, to many more pages in a full report. Generally, these formats are written in a less academic style than research articles or scientific reports, with shorter sentences, less jargon, and few formal citations.

Color

Posters, chartbooks, and briefs are often printed in two or three colors to enhance their appeal and to convey information such as differences across groups or statistical significance of findings. However, they are frequently photocopied into black and white, whether for a grayscale handout based on a poster or chartbooks, or for broader distribution of briefs. Plan for this eventuality by designing these documents so they can be interpreted in black and white. See "Use of Color" in chapter 7 for additional suggestions.

COMMON FORMATS FOR APPLIED AUDIENCES

Common formats for presenting statistical results to applied audiences include posters, policy or issue briefs, chartbooks, and general-interest articles. Executive summaries are often included at the beginning of reports and chartbooks. Below I describe the audiences, contents, and layouts for each of these formats, with detailed suggestions for posters and issue briefs. I then show how chartbooks, general-interest articles, and executive summaries can be created by adapting elements of the other formats. In the closing section, I explain how to assess which format is best suited to your specific audience and objectives. Examples from a variety of topics and types of studies illustrate how these documents can be used to report elementary statistics and reference data as well as results of more complex statistical analyses. The length, structure, and contents of each format vary depending on the client, conference, or publication, so check the applicable guidelines before writing.

Posters

An assortment of posters is a common way to present results to viewers at a professional conference. Posters are a hybrid form—more detailed than a speech but less than a paper, more interactive than either. Different people will ask about different facets of your research. Some may conduct research on a similar topic or with related data or methods. Others will have ideas about how to apply or extend your work, raising new questions or suggesting other contrasts, ways of classifying data, or presenting results. In addition, presenting a poster provides excellent practice in explaining quickly and clearly why your research is important and what it means—a useful skill to apply when revising a speech or paper on the same topic. See appendix B for a comparison of the content, formatting, and audience interaction for a paper, speech, and poster about the same research project.

By the end of an active poster session, you might have learned as much from your viewers as they have from you, especially if the topic, methods, or audience are new to you. For example, at David Snowdon's first poster presentation on educational attainment and longevity using data from the Nun Study, another researcher returned several times to talk with Snowdon, eventually suggesting that he extend his research to focus on Alzheimer's disease, which lead to an important new direction in his research (Snowdon 2001).

AUDIENCES FOR POSTERS

Preparing a poster means more than simply printing out pages to be tacked onto a bulletin board in a conference hall. It also involves writing an associated narrative and handouts, and preparing short answers to likely questions, all of which should be adapted to the audience. In contrast to chartbooks and issue briefs, which are generally written for nonstatistical audiences, posters are used for both statistical and nonstatistical audiences. For instance, the annual meeting of the American Public Health Association draws both academics who conduct statistical analyses and public health practitioners who typically do not. In such situations, use nontechnical vocabulary, examples, types of charts, and means of presenting results of statistical tests on the poster, saving the methodological details and statistical tables for handouts (see below).

CONTENTS AND ORGANIZATION OF A POSTER

Research posters are organized like scientific papers, with separate sections devoted to the background, objectives, data and methods, results, and conclusions. Because viewers read the posters at their own pace and at close range, more detail can be included than in slides for a speech, but less detail than in a full written document. Be selective, concentrating on one or two issues in the poster. Do not simply post pages from the full paper. Adapt them, using the formatting ideas listed below under "Other Design Considerations." See Miller (2007b), Briscoe (1996), or Davis (1997) for more recommendations about designing research posters.

To determine how many pages[1] you have to work with, find out the dimensions of your assigned space and design your poster to fit that space. A trifold tabletop presentation board (3' high by 4' wide) will hold roughly a dozen 8.5" by 11" pages, organized into three panels. The left- and right-hand panels each hold about three pages, while the wider middle panel can accommodate another half-dozen. Figure 13.2 shows a suggested layout.

- In the left-hand panel, set the stage for the research question, stating why the topic is important, summarizing major empirical or theoretical work on related topics, and stating your hypotheses or research questions, as in the material shown in figures 12.1, 12.2, 12.4, and 12.7. Include a one-page abstract of your project.
- In the middle panel, briefly describe your data source, variables, and methods as in figures 12.5 and 12.6, then present results in tables, charts, or bulleted text, using charts like those in figures 12.8 and 12.9. Alternatively, enlarge the type size on table 11.1, display the statistically significant results in color in lieu of standard errors, test statistics, or p-values, and accompany the table with bulleted annotations.
- In the right-hand panel, summarize your findings and relate them back to the research question or project aims, discuss strengths and limitations of your approach, identify research or policy implications, and suggest directions for future research. For instance, write text bullets to convey material about the birth weight study from boxes 10.3 and 11.3.

An 8' by 4' bulletin board can accommodate several additional pages, allowing you to go into somewhat more depth and to present

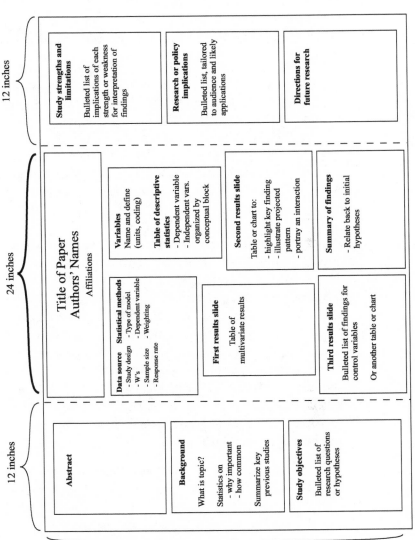

Figure 13.2. Example layout for a research poster

results of additional analyses. See Miller (2007b) for an example layout.

OTHER DESIGN CONSIDERATIONS

A few other issues to keep in mind as you design your poster pages.

- Write a short, specific title that fits on one or two lines in a large type size. The title will be potential readers' first (and sometimes only) glimpse of your poster, so make it interesting and easy to read from a distance—at least 48 point, ideally larger.
- Print all text, tables, and chart labels in at least 14-point type to enhance ease of reading. Take advantage of the smaller type size by combining the contents of two slides designed for a speech onto a single page. For example, the descriptions of data and variables from figures 12.5 and 12.6 could comprise one poster page.
- Modify tables and charts to simplify presentation of statistical test results (see "Adapting Tables and Charts for Slides" in chapter 12).
- Make judicious use of color to differentiate among groups in charts or to highlight statistically significant results in tables. Use a clear, white, or pastel background, with most text in black or another dark color, and a bright, contrasting shade for elements you wish to emphasize.
- If there is space, accompany charts and tables with a narrative explanation. Use a "chartbook layout" such as figure 12.2, with a chart or simple table on one side of a landscape page, accompanied by bulleted annotations on the other.

Use presentation software to create your pages or adapt them from related slides, facilitating good page layout with adequate type size, bulleted text, and page titles. Such software also makes it easy to create matching handouts (see "Handouts to Accompany a Poster" below and "Handouts" in chapter 12).

ORAL INTRODUCTION TO A POSTER

Prepare a brief overview to introduce the purpose, findings, and implications of your work. Keep it short—a few sentences that highlight what you are studying and why it is important. After hearing your introduction, listeners will either nod and move along, or comment on some aspect of your work that intrigues them. You can then tailor additional discussion to the individual listener, adjusting the focus and amount of detail to suit his or her interests. Gesture at the pertinent slides as you make each point, using "Vanna White" scripts to introduce and explain your tables or charts (see "Explaining a Chart 'Live'" in chapter 12). Also prepare short answers to likely questions about various aspects of your project, such as why the work is important from a research or policy perspective, or descriptions of data, methods, and specific results. Think of these as modules—succinct descriptions of particular elements of your research that you can choose among in response to questions that arise. Finally, consider writing a set of questions to ask your visitors, which can help you engage them during the session, and improve or extend your project afterward. These might include questions about their reactions to your findings, relevant literature, data, and how they might apply your findings to their work.

HANDOUTS TO ACCOMPANY A POSTER

For conference presentations, prepare handouts to distribute to interested viewers. Handouts can be created from slides printed several to a page by presentation software, along with a cover page containing the title, abstract, and your contact information. Or package an executive summary or abstract with a few key tables or charts.

Issue and Policy Briefs

Issue and policy briefs are just that—short summaries of how your research findings apply to some real-world issue or policy. Often structured around a set of questions, they are intended for legislators, advocates, and others interested in your topic but not the technical details of your statistical analyses. See Musso et al. (2000) for additional guidelines about writing briefs, DiFranza et al. (1996) for pointers on communicating policy-related research to the media and other lay audiences.

Many issues and policies are of interest to a variety of audiences or stakeholders who represent a range of perspectives and potential applications of your research findings. As you write your brief, put yourself in the shoes of each likely audience and explain how your questions and findings apply to them. For instance, in their series of issue briefs on children's mental health, Warner and Pottick (2004) identify parents and other caregivers, service providers, and policy makers as parties who are likely to be interested in their findings, then describe how each group might best respond to the study findings (box 13.1). If your topic is relevant to several diverse audiences, consider writing different briefs that address their respective interests and viewpoints.

CONTENTS OF BRIEFS

Introduce the topic and why it is of concern to the intended audience. Don't make them decipher for themselves how your analyses fit their questions. Instead, *you* figure it out before you write, then explain accordingly. Often, applied readers' questions will affect how you code your data or the types of analyses you conduct, so plan ahead by familiarizing yourself with their interests and likely applications of your research results; see "Substantive Context" in chapter 8.

Title

Write a title that convinces potential readers that your brief merits their attention. Make it like a newspaper headline—informative and enticing—and incorporate the key question or main conclusions of your study.

- Instead of "Spending Patterns of People with and without Health Insurance," write "What Do People Buy When They Don't Buy Health Insurance?" (Levy and DeLeire 2002).
- Instead of "Prevalence of Multiple Mental Health Problems among Children," write "More than 380,000 Children Diagnosed with Multiple Mental Health Problems" (Warner and Pottick 2004).

From "More than 380,000 Children Diagnosed with Multiple Mental Health Problems" (Warner and Pottick 2004).

"HOW WE SHOULD RESPOND TO THE FINDINGS"

"Preventing multiple psychiatric problems may be as important as treating them. Prevention means focusing on why co-occurring disorders develop and eliminating the factors that put the children at risk for them. If multiple problems can't be prevented, they should be detected early and treated promptly in order to minimize the substantial burden of psychiatric illness, and encourage positive outcomes.

"Parents and other caregivers are in the best position to observe a child's symptoms and describe them in detail to service providers so that diagnoses are accurate. They should advocate vigorously for their child, making sure that the assessments are thorough, and that treatment reflects the latest research on co-occurring disorders.

"Service providers must stay up-to-date with research on how co-occurring disorders develop and the best ways to treat them. They should also offer support programs for families and caregivers to help them respond to the special needs of these children and to reduce their risk of developing chronic mental illness as adults.

"Policymakers should encourage clinical trials that include children with multiple diagnoses and ensure that community-based mental health programs offer child psychiatrists who are expert in co-occurring disorders."

If you report a numeric finding in your title, keep in mind that readers may latch onto it as a "factoid" to summarize your conclusions, so select and phrase it carefully (McDonough 2000).

Briefs are often organized around a series of questions that your audience is likely to ask, with results of your study interpreted to answer those questions. Again, those familiar W's are a handy checklist for identifying relevant questions and organizing your brief:

- Why is the issue important?
- What were your main findings?
- How many people or institutions or countries are affected? Who and where?
- How do your findings apply to the issue or policy at hand?
- How do they correspond to proposed or existing policies? What modifications or new solutions do your results suggest?

Create a paragraph heading for each key question, then present the associated findings in a simple chart, accompanied by short, straightforward descriptions. Limit your design to no more than two charts per page. For some questions, incorporate a few numeric facts in lieu of a chart, providing enough background information that readers can interpret those facts.

Sidebars

Incorporate basic contextual information (W's) into the narrative. If additional information on the data or methods is essential for understanding the material in the brief, include it in a sidebar and provide a citation to a published paper or report where interested readers can find the technical details. Avoid a lengthy description of methods. For sidebars about multivariate statistical methods, see Miller (2013a), chapter 20.

Glossary

In some cases, you will intentionally use technical terms in an issue brief to tie your findings to other similar works or if there is no suitable everyday synonym. For instance, perhaps you use a specific measurement technique that is widely used in the field and should be referred to by its usual name. If you use such terms in your brief, provide a glossary or short definitional note. For instance, in their issue brief on the prevalence of multiple mental health problems among children, Warner and Pottick (2004) include the following note defining the "GAF" (Global Assessment of Functioning), which they refer to in their issue brief by that acronym because of its widespread use in the field.

> "*Note*: GAF is a standardized diagnostic tool used to determine the severity of emotional disturbances. A score of 50 or below means that such children have a moderate degree of interference in functioning in most social areas, or a severe impairment of functioning in one area."

Limit the glossary to no more than a handful of terms, replacing other technical language with more familiar wording in the body of the text so that definitions are not needed.

LENGTH AND FORMAT

As their name suggests, issue and policy briefs are generally very short. They are intended for people who can't devote much time to reading about any one topic, often because they must familiarize themselves with many different subjects. A 2001 survey of government policy makers showed that they prefer summaries of research to be written so they can immediately see how the findings relate to issues currently facing their constituencies, without wading through a formal research paper (Sorian and Baugh 2002). Complaints that surfaced about many research reports included that they were "too long, dense, or detailed," or "too theoretical, technical, or jargony." On average, respondents said they read only about a quarter of the material they receive for detail, skim about half of it, and never get to the rest.

To ensure that yours is among the material they read and remember, keep it short and specific, explaining the major questions and answers in plain language, and using charts or bullets to highlight major findings. Common styles for briefs include one-page memos, those that cover both sides of a single page, or simple bifold (e.g., four-page) or trifold (six-page) documents.

Figure 13.3 shows a sample layout for a two-page issue brief, modeled after those designed and written by Pottick and Warner (2002; Warner and Pottick 2004) with example titles and section headings and suggested placement of elements such as the sidebar and glossary or vocabulary note. Positioning of charts may vary from the design shown here. For example, charts might appear on both the

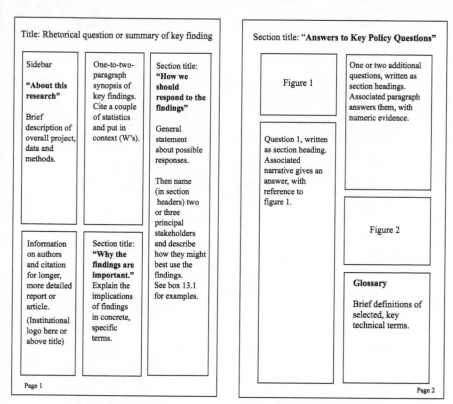

Title: Rhetorical question or summary of key finding

Sidebar **"About this research"** Brief description of overall project, data and methods.	One-to-two-paragraph synopsis of key findings. Cite a couple of statistics and put in context (W's).	Section title: **"How we should respond to the findings"** General statement about possible responses. Then name (in section headers) two or three principal stakeholders and describe how they might best use the findings. See box 13.1 for examples.
Information on authors and citation for longer, more detailed report or article. (Institutional logo here or above title)	Section title: **"Why the findings are important."** Explain the implications of findings in concrete, specific terms.	

Page 1

Section title: **"Answers to Key Policy Questions"**

Figure 1

Question 1, written as section heading. Associated narrative gives an answer, with reference to figure 1.

One or two additional questions, written as section headings. Associated paragraph answers them, with numeric evidence.

Figure 2

Glossary

Brief definitions of selected, key technical terms.

Page 2

Front page of brief   **Back page of brief**

Figure 13.3. Example layout for a two-page issue brief
Note: Text shown on the diagram in bold type with quotation marks would be included verbatim in the issue brief. Other text on the diagram gives guidelines about general topics to include in the brief.

first and second pages, next to the accompanying descriptions. Longer briefs can accommodate more findings, each described in a short paragraph and identified by a clear subheading or question. If space permits, leave a blank area for a mailing label to facilitate distribution. For briefs that are part of a series, add a banner for the series title.

Chartbooks

Chartbooks present numeric information in a format that is easily accessible to a wide range of audiences. Longer than issue briefs, they accommodate more detailed results on a given topic, or findings on a wider range of topics within a single document. For example, the National Center for Health Statistics produces an annual chartbook titled "Health, United States" summarizing a variety of health status, health behaviors, and access to health care indicators (National Center for Health Statistics 2013). The Social Security Administration publishes annual chartbooks that describe income levels and sources among the elderly in the United States (US SSA 2012).

CONTENTS

Begin the chartbook with a brief narrative introduction to the purpose and methods of the work, then present results in a series of charts, each addressing one aspect of the topic. Place each chart on a separate page, accompanied by two or three text bullets summarizing the patterns. Follow the results section with a short summary of findings and discussion of policy implications and directions for future work, using either a bulleted list or paragraph format. For readers interested in the technical details, include an appendix at the end of the chartbook or cite the associated statistical paper.

LAYOUT AND FORMAT

The design of each chartbook page is similar to the design of a slide for a speech, and follows many of the same guidelines for an effective title and length and organization of text bullets. Type size can be smaller than that in slides (e.g., 12- to 14-point type), accommodating more information on each page. A chartbook about the birth weight study might include an executive summary like that in box 13.2 (below), a short description of data and methods (figure 13.4), and results presented in charts like those in chapter 12. For simple charts, provide annotation on the same page as the chart (e.g., figure 12.2). For charts that require a full page, place accompanying annotations on a facing page. As you design the chartbook, include and explain reference lines, bars, or other elements to help readers interpret the values you show. See comment 6 in box 13.3 below for an illustration.

Data and Methods

Data sources

- 1988–1994 National Health and Nutrition Examination Survey (NHANES III)
  - Nationally representative sample of US population
    - Oversample of Mexican Americans
  - Cross-sectional
  - Population-based

- $N = 9,813$
  - 3,733 non-Hispanic white
  - 3,112 Mexican American
  - 2,968 non-Hispanic black

- Information on all variables used here taken from household survey portion of the study

Statistical methods

- Chi-square tests for bivariate associations between categorical independent variables and LBW

- t-tests for differences in means of continuous birth weight by categorical independent variables

- Weighted to national level with sampling weights from NHANES III
  - Corrected for complex survey design

Figure 13.4. Example data and methods summary for a chartbook

Descriptive Reports

Descriptive reports or those that serve as reference data sources are often formatted using a variant of a chartbook layout but including both tables and charts, and replacing text bullets with paragraphs. For example, the Department of Health and Human Services' periodic report to Congress on indicators of welfare dependence (US DHHS, Office of Human Service Policy 2013) divides the information into major topic areas, each of which comprises a section of the report. Start each section with a two- or three-page prose introduction to the measures, data sources, and questions covered in that section, then present tables and charts, each annotated with several text bullets or a brief paragraph describing the patterns. Use tables to report precise numeric values, charts to portray approximate composition, levels, or trends in the outcomes under study.

Executive Summaries

Executive summaries are one- to two-page synopses of a study that are often included at the beginning of a report or chartbook. They give an overview of the key objectives, methods, findings, and implications of the work that can be read and understood in a few minutes by busy executives or others who must digest the main points from many documents quickly. They are similar in general content and structure to abstracts, but emphasize the questions and answers of the study, with less detail on methods or statistical test results. To make the information easily accessible, executive summaries are often written in a bulleted format with short, simple sentences (box 13.2). Choose the numbers you report carefully; they are often chosen as "sound bites" to characterize conclusions of the entire work.

General-Interest Articles

General-interest articles typically aren't divided into the formal parts that characterize a scientific paper. Instead, write a coherent, logical story line with numeric facts or patterns as evidence, incorporating information about context, data, and methods into the body of the narrative. In multipage articles, consider using subheadings to guide readers through the different topics within your work.

In general-interest articles, numeric facts are often used to highlight the importance of the topic, to provide other background information, or to illustrate patterns of distribution or association. If you include charts or tables, keep them simple and focused, then refer to them by name or number as you describe and interpret a sample number or two in the text or a sidebar. To adapt sentences involving citations to other work for a lay audience, replace the citation with the name of the author or source.

To illustrate, box 13.3 shows an excerpt from a two-page article in the *New York Times* about the physical impact of the planes that hit the Twin Towers on September 11, 2001 (Lipton and Glanz 2002). The article includes some fairly technical information, but is written for an educated lay audience. I have accompanied the excerpt with the figure (13.5) from their article, and annotated it to show how the authors included explanations, charts, and analogies to elucidate numbers and concepts in the article. Some explanations define terms (joules) and concepts (how speed relates to energy on impact); others provide comparisons to clarify what the numbers mean or why certain cutoffs are pertinent. Without those elements, readers who aren't

BOX 13.2. EXECUTIVE SUMMARY

BACKGROUND

- Low birth weight (LBW) is a widely recognized risk factor for infant mortality and poor child health.
- Although only 7.5% of all births are LBW, such infants account for more than 75% of infant deaths.
- Black infants weigh on average about 260 grams less than non-Hispanic white infants. Rates of LBW among non-Hispanic black infants are approximately twice as high as for non-Hispanic white infants (13.0% and 6.5% in 2000).
- Low socioeconomic status (SES) is also associated with lower mean birth weight.

STUDY OBJECTIVE

- To assess whether the lower average SES of non-Hispanic black infants in the United States explains why they have lower mean birth weight than non-Hispanic white infants.

DATA AND METHODS

- Data on 9,813 children were taken from the 1988–1994 National Health and Nutrition Examination Survey (NHANES III)—a cross-sectional, population-based sample survey of the United States.
- Birth weight data were collected in parental interviews about children aged 10 or younger at the time of the survey.
- To assess the extent to which SES explained observed racial differences in birth weight, mean birth weight for each racial/ethnic group was calculated within socioeconomic strata defined by mother's education in order.

KEY FINDINGS

- Regardless of race, children born into low SES families have lower mean birth weight than those born at higher SES.
- At each socioeconomic level, non-Hispanic black infants

weighed 180 to 225 grams less than non-Hispanic whites or Mexican Americans.

CONCLUSIONS
- Further research is needed to investigate possible reasons for the birth weight deficit among non-Hispanic black compared to other infants. These include:
  - less access to health care,
  - higher rates of poor health behaviors,
  - greater social stress, and
  - intergenerational transmission of health disadvantage.

familiar with physics, engineering, or airline regulations would find it hard to grasp the purpose or interpretation of the numbers in the article. In each comment, I also identify which principle or tool the authors used in the associated sentence. Note that the authors did not name the various principles and tools as they used them, but simply integrated them into the narrative.

Impact speed of 9/11 flights and comparison speeds

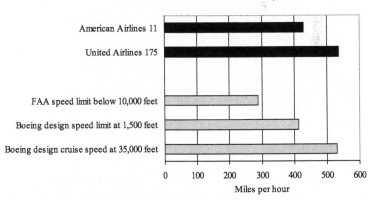

Figure 13.5. Bar chart to accompany a written comparison
Source: Lipton and Glanz 2002
Note: FAA = Federal Aviation Administration

". . . (1) The government's analysis put the speeds [on impact] at 586 m.p.h. for the United flight and 494 m.p.h. for the American one. In both cases, (2) the planes were flying much faster than they should have been at that altitude: The aviation agency's limit below 10,000 feet is 287 m.p.h. . . .

"(3) The energy of motion carried by any object, called the kinetic energy, varies as the square of its velocity, so even modest differences in speed can translate into large variations in what the building had to absorb. (4) That means that while the United jet was traveling only about a quarter faster than the American jet, it would have released about 50% more energy on impact . . . (5) Even at a speed of only about 500 m.p.h., a partly loaded Boeing 767 weighing 132 tons would have created about three billion joules of energy at impact, the equivalent of three-quarters of a ton of T.N.T."

After (6) presenting the respective speeds in a simple bar chart (figure 13.5), the authors explain that (7) both jets were traveling at speeds that exceeded the Boeing design speed limit at 1,500 feet, and the United jet exceeded even the design cruise speed at 35,000 feet. Such speeds threatened the structural integrity of the planes even before they struck the buildings, because "(8) The lower the plane goes, . . . the thicker the air becomes, so the slower the plane must travel to avoid excessive stress."

COMMENTS

(1) Reports the actual speeds of each plane. *Basic principle: Report numbers.*

(2) Compares the planes' speeds with a cutoff: the FAA limit for flights at that altitude. *Basic principle: Compare against a standard to help interpret numbers.*

(3) Explains in lay terms how differences in speed translate into differences in energy on impact, paraphrasing the meaning of

"kinetic energy." *Basic principle: Define concepts using simple wording, and avoid or paraphrase jargon.*

(4) Reports results of calculations to illustrate how much more energy the second plane released, applying the general formula given in the preceding sentence. *Basic principle: Report numbers and results of calculations. Tools: Ratio and percentage difference.*

(5) Reports results of calculation of absolute energy generated by the second plane. Units of measurement (joules) are compared against a more familiar quantity. *Basic principle: Interpret numbers using analogies or metaphors.*

(6) Uses a bar chart to illustrate speeds of the two planes and how they compare to industry design speeds. *Basic principle: Choose the right tools.*

(7) Explains that the planes were traveling too fast for conditions by comparing their speeds against the design speed limit at both the altitude where the planes were flying and cruise altitude. *Basic principle: Compare against meaningful cutoffs.*

(8) Uses colloquial language to explain the physical principles why the design speed is slower for lower altitudes. *Basic principle: Explain complicated concepts in everyday language.*

SUMMARY

The types of documents used to present numeric results to applied audiences all share certain attributes that help make the information accessible and useful to nonstatisticians.

- They emphasize substantive questions and answers over statistical methods and findings.
- They replace technical terms with their colloquial equivalents.
- They use simplified charts or tables to present numeric findings and associated statistical test results.

- They include limited (if any) technical information on statistical methods and data, placing that material in appendixes (for descriptive reports or chartbooks) or sidebars (for briefs), and referring to the associated scientific article for details.

Exceptions to these generalizations include posters for a research audience and analytic reports, in which data and methods are described in the body of the work and results are presented with technical detail similar to that in a scientific paper.

The different formats described in this chapter vary substantially in terms of the audiences and objectives for which they are best suited. In some instances, one of these formats will work best:

- Issue briefs or policy briefs, which are written for applied audiences who need to see what your findings mean for an issue or policy of interest to them without reading a long, detailed statistical report.
- General-interest articles, which are written for lay audiences, with a more essay-like structure, few citations, and descriptions of a handful of numeric facts or patterns.

In other instances, you will choose among two similar variants:

- Posters and speeches, which are visual and spoken versions of the same material, and can be used for similar audiences at professional conferences; see appendix B.
- Chartbooks and descriptive reports, which differ principally in whether they illustrate general patterns (charts in either format) or present precise values (tables in reports). Reports also typically include more prose than do chartbooks, using full paragraphs in place of bulleted text. Both often include executive summaries.

CHECKLIST FOR WRITING FOR APPLIED AUDIENCES

- When writing for an applied audience, determine which format is most appropriate, taking into account whether readers need or want input into the analytic methods, assumptions, or variables you will use,

the level of statistical training among expected readers,

the likely application of results, and

the amount of time readers have to digest findings.

- Integrate the W's into your narrative.
- Explain briefly how key variables were measured, if this affects interpretation of your findings.
- Emphasize questions and answers rather than statistical methods.
- Describe direction and size of patterns, using approaches described in chapters 2, and 9, and wording for results of calculations explained in chapter 5.
- Replace technical language with familiar synonyms or analogies that convey the underlying concepts. Incorporate definitions into the description of findings, or provide a glossary or vocabulary note.
- Simplify charts and tables to focus on one pattern at a time and to emphasize key patterns only.
- Adapt your presentation of statistical results, using formatting such as color, italics, or boldface to convey results of statistical tests.

APPENDIX A

IMPLEMENTING "GENERALIZATION, EXAMPLE, EXCEPTIONS" (GEE)

One of the basic principles for describing a relationship among two or more variables is to summarize, characterizing that association with one or two broad patterns. In chapter 2, I introduced a mantra, GEE, for "generalization, example, exceptions," to use as a guide on how to write an effective summary. Generalize by stepping back to look at the forest, not the individual trees, describing the broad pattern rather than reporting every component number. Illustrate with representative numbers to portray that general pattern. Finally, if the general pattern doesn't fit all your data, identify and portray the exceptions. For inexperienced writers, this can seem like a daunting task.

In this appendix, I explain six steps to guide you through implementing a GEE. See online materials for a demonstration. After creating a chart and table to present data on the pattern, proceed through several intermediate steps to identify and characterize the patterns for your final written description. The notes, calculations, and scribbles generated in those steps will not appear in the final written narrative, but are an important part of the analytic process involved in writing a succinct but thorough summary.

STEP 1: DISPLAY THE DATA IN A TABLE AND A CHART

Even if you plan to use a table or prose in your document, the chart version may help you see patterns in your data more easily as you write your GEE. It doesn't have to be pretty—even a hand-drawn version will work fine for this purpose—as long as it is an appropriate type of chart for the task (see table 7.1), is drawn to scale, and is labeled well enough that you can recognize the variables and assess approximate numeric values.

In both table and chart, organize nominal variables or items in a logical order, using empirical or theoretical criteria (see "Organizing Charts to Coordinate with Your Writing" in chapter 7 or Miller 2007a), facilitating an orderly comparison in the subsequent steps of the GEE. For instance, the AIDS questions in table A.1 are sorted in descending order of percentage correct in the modal language group (English), making it easy to identify the topics with the highest values.

STEP 2: IDENTIFY THE DIMENSIONS OF THE COMPARISON

Identify the dimensions of the comparison—one for each variable or set of variables in your table or chart. In an interaction (chapter 9), each predictor variable is one dimension of the comparison. In a table, the rows and columns each comprise one dimension; panels of rows or spanners across columns often indicate the presence of additional dimensions. In a chart, the axes and legends each comprise one dimension. Vary only one dimension of the chart or table at a time, keeping the others constant. In a multiple-line trend chart like figure A.1, there are two separate comparisons:

(1) *Moving left to right along one line.* In figure A.1, this comparison shows how the variable on the y axis (price of housing) varies with time (the x variable), within one region (value of the z variable, shown in the legend).

(2) *Moving (vertically) across lines.* In figure A.1, this comparison shows how price (the y variable) varies across regions (the z variable), at one time point (value of the x variable).

Median sales price of new single-family homes, by region, United States, 1980–2000

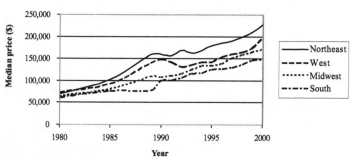

Figure A.1. Generalizing patterns from a multiple-line trend chart
Source: US Census Bureau 2001a

In a three-way table, there are two comparisons:

(1) *Moving down the rows within one column.* Table A.1 shows how AIDS knowledge (in the interior cells) varies by row (topic) within one language group (column).

(2) *Moving across the columns within one row.* Table A.1 shows how AIDS knowledge varies by column (language of respondent) for one topic (row).

STEP 3: IDENTIFY A REPRESENTATIVE EXAMPLE

Having noted each of the dimensions of comparison, identify one representative example as the basis for each generalization about the shape of the pattern.

- For a comparison across a series of related outcomes, a good starting point is a summary measure (e.g., in table A.1, the mean percentage of "likely" questions correct) that combines results for the various component variables. Lacking a summary measure, pick a value of particular interest or start at one end of the axis, column, or row (the best-answered or worst-answered topic, in table A.1).

- For a comparison across groups, a good starting point for a representative value is the overall sample (e.g., all language groups combined). Alternatively, use the modal (most common) group—English speakers, in table A.1—or a group of particular interest for your research question.

Follow steps 4 and 5 to ensure that your example is in fact representative of a general pattern. If not (e.g., if it turns out to be an exception), try step 3 again with a different example until you've found one that is generalizable.

STEP 4: CHARACTERIZE THE PATTERN

Using your example value, describe the shape of the pattern, including direction, magnitude, and, for a scientific audience, statistical significance. Make notes in the margins of your table or chart or on an accompanying page. Don't worry about writing complete sentences

Table A.1. Generalizing patterns within a three-way table

Percentage of respondents answering AIDS transmission questions correctly, by language spoken at home and language used on the questionnaire, New Jersey, 1998

Mode of transmission	Language spoken at home and language used on questionnaire		
	English (N = 408)	Spanish/ English ques. (N = 32)	Spanish/ Spanish ques. (N = 20)
Likely modes of transmission			
Sexual intercourse with an infected person	93.6	87.5	95.0
Sharing needles for IV drug use*	92.4	90.6	65.0
Pregnant mother to baby*	89.5	75.0	80.0
Blood transfusion from infected person*	87.5	81.3	60.0
Mean percentage of "likely" questions correct*	91.7	83.6	75.0

Source: Miller 2000a.
* Difference across language groups is significant at $p < 0.05$.

at this stage. Abbreviate concepts to use as a basis for your written description with short phrases, upward- or downward-pointing arrows, and <, =, or > to show how values on different categories, topics, or time points relate to one another.

Direction

For *trends* across values of an ordinal, interval, or ratio variable such as time, age, or price, describe whether the pattern is

- level (constant) or changing;
- linear (rising or falling at a steady rate), accelerating, or decelerating;
- monotonic or with a change of direction (such as a notable "blip" or other sudden change).

For *differentials* across categories of a nominal variable such as geographic region, language, or gender, indicate which categories

Generalization 1: Down the rows. On the summary measure of knowledge of "likely" modes of AIDS transmission, English speakers score higher than Spanish/English speakers, who in turn score higher than Spanish/Spanish speakers. The difference across language groups in mean percentage of "likely" questions correct is statistically significant, as indicated by the asterisk at the end of the row label (see note to table).

Check: Does the pattern in the row showing mean percentage correct apply to the other rows? In other words, does the generalization from the summary row fit each of the component questions?

Answer: The generalization fits all but the sexual intercourse and mother-to-baby questions. In addition, the difference across language groups in knowledge of sexual intercourse as a likely means of AIDS transmission is not statistically significant. Hence those questions are exceptions to the general pattern, in terms of both direction and statistical significance.

Generalization 2: Across the columns. Among English speakers, the best understood "likely" AIDS transmission topic was transmission via sexual intercourse, followed by sharing IV needles, transmission from pregnant mother to baby, and blood transfusion. (As noted above, the question topics were arranged in the table in descending order of correct answers for the English-speaking group, facilitating this description.)

Check: Does this same rank order of topics apply to the other language groups? In other words, does the generalization from the modal language column also describe each of the other columns?

Answer: Spanish/English speakers did best on the needles question and least well on mother to baby. Among Spanish/Spanish speakers, the rank order of the two middle questions is reversed. In this analysis, number of Spanish speakers is small, so these exceptions would not be emphasized.

have the highest and lowest values and where other categories of interest fall relative to those extremes, as explained in chapter 9.

Magnitude

Use one or two types of quantitative comparisons (chapter 5) to calculate the size of the trend or differential. If the calculations involve only a few numbers and basic arithmetic (e.g., a ratio of two numbers, or a percentage change), include those calculations in your working notes, including units. For more complex or repetitive calculations, such as confidence intervals for each of a dozen independent variables, save your work in a spreadsheet, then annotate it to indicate which calculations correspond to which aspects of the GEE for your own future reference.

Scribble down descriptive words or phrases to depict the size of the variation. Is the trend steep or shallow? Is the differential marked or minuscule?

Statistical Significance

Note patterns of statistical significance on your table or chart, particularly if it does not include symbols to indicate which results are statistically significant. Are most of the associations in the table statistically significant? If so, generalize that finding. If most are not, the lack of statistical significance is your generalization. Finally, if only some portions of your table or chart have statistically significant findings, try to identify what they have in common so you can summarize the patterns to the extent possible.

STEP 5: IDENTIFY EXCEPTIONS

If parts of your table or chart depart appreciably from the generalization you have made in the steps above, they are exceptions. As noted in chapter 9, exceptions come in three flavors: direction, magnitude, and statistical significance. A few more illustrations:

Exceptions in Direction

In figure A.1, median sales prices dipped in the early 1990s in the West, but continued upward or remained level in the other regions—an example of a different direction of trend. The West was the exception.

From 1981 onward, the Northeast had the highest prices ("Northeast > West"). In 1980, sales prices in the Northeast were slightly below those in the West ("Northeast < West"), an example of a contrasting direction of a cross-sectional comparison. The year 1980 was the exception.

Exceptions in Magnitude

In figure A.2a, the odds ratio of emergency room use by racial group is much larger among the non-poor than in the other two income groups (compare brackets 2 and 3 to bracket 4). Generalize based on the two income groups for which the racial gap in ER use is similar (poor and near poor), and then point out that the non-poor are the exception.

Exceptions in Statistical Significance

In table A.1, the sexual intercourse question is the only one for which the language difference in AIDS knowledge is not statistically

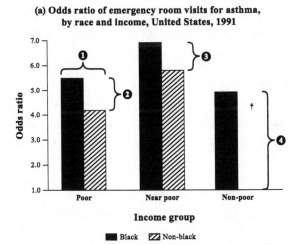

(a) Odds ratio of emergency room visits for asthma, by race and income, United States, 1991

Figure A.2a. Generalizing one pattern within a three-way chart: Within clusters
Source: Miller 2000b.
Notes: Taking into account mother's age, educational attainment, and marital history; number of siblings; presence of smokers in the household; low birth weight (<2,500 grams) and preterm birth (<37 weeks' gestation). Compared to non-black non-poor.
† Difference across racial groups significant at $p < 0.05$ for non-poor only.

significant. For that table, statistical significance is the rule (generalization), and lack of statistical significance is the exception.

On the printed copy of your table or chart, circle or otherwise mark exceptions to your general pattern. If your table or chart is complicated, use a colored highlighter to shade which parts share a common pattern and which deviate from that pattern.

STEP 6: WRITE THE DESCRIPTION

Working from your notes and calculations, write a systematic description of the patterns. For relationships among three or more variables, organize the GEE into one paragraph for each type of comparison—e.g., one for the pattern "across the columns," another for "down the rows"; see "Step 2: Identify the Dimensions of the Comparison" (above).

Start each paragraph with a topic sentence that identifies the main concepts or variables in the comparison. Provide a verbal sketch of the general pattern, selecting verbs and adjectives to convey direction and magnitude. Follow with one or more sentences illustrating that generalization with results of your calculations, reporting the numbers from which they were calculated or (if many numbers are involved), referring to the associated table or chart. Finally, describe and

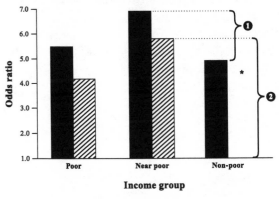

(a) Odds ratio of emergency room visits for asthma, by race and income, United States, 1991

Income group

■ Black ⧄ Non-black

Figure A.2b. Generalizing a second pattern within a three-way chart: Across clusters

Note: Taking into account mother's age, educational attainment, and marital history; number of siblings; presence of smokers in the household; low birth weight (<2,500 grams) and preterm birth (<37 weeks' gestation). Compared to non-black non-poor

* Difference across income groups significant at $p < 0.05$ for non-blacks only.

document any exceptions. See "'Generalization, Example, Exceptions' Revisited" in chapter 9 for suggested wording to differentiate general patterns from exceptions.

Figures A.2a and b, table A.1, and the associated text boxes illustrate how to identify the dimensions, select a starting point for each generalization, and test the generalization for exceptions of direction and magnitude.

APPENDIX B
COMPARISON OF RESEARCH PAPERS, ORAL PRESENTATIONS, AND POSTERS

Table B.1. Comparison of research papers, oral presentations, and posters—materials and audience interaction

	Research paper (chaps. 10 and 11)	20-min. oral presentation (chap. 12)	4' by 8' poster presentation (chap. 13)
Written materials			
Form	Printed document (paper or web)	Slides or overhead transparencies	Poster pages or single-sheet poster
Length	• 12–16 pages (research brief) • 10–25 pages (full length article); see instructions to authors for specific journal	• Average of 1 slide/minute • Less time for text slides • More time for table/chart slides	• 12 pages (tri-fold table-top presentation board) • 15–20 pages (4' by 8' bulletin board)
Style	Full sentences and paragraphs; formal essay structure	• Bulleted phrases replace full sentences • Charts and tables simplified from research paper version	• Full paragraphs in abstract • Bulleted sentences in introduction and conclusion • Bulleted phrases replace full sentences in data & methods & results • Charts and tables simplified from research paper version

Type size	12-point	• Slides titles: 24 point or larger • Text: 18 point or larger	• Poster title: 40 point or larger • Page titles: 20 point or larger • Text: 14–16 point
Color	Black type on white background	Dark background with light main text *or* pale background with dark text Contrasting color for emphasis (e.g., statistical significance)	Clear or pale colored background Dark color for most text Contrasting color for emphasis (e.g., statistical significance)
Other formatting	Use subheadings to guide readers within sections (e.g., naming subtopics within the introduction, literature review, or results)	Use slide titles to guide viewers, integrating specific topic, purpose, or finding of each slide	Use page titles to guide viewers, integrating specific topic, purpose, or finding of each page
Spoken materials	Not applicable	Speaker's notes: Either full narrative script or selected reminders, coordinated with slides. Include "Vanna White" notes for yourself as needed to describe charts, tables, or other diagrams.	A few sentences summarizing main objectives and findings, spoken to each viewer Separate 1–2 minute modules for: background, literature review, data and methods, each major set of results, conclusions, research.

(continued)

Table B.1. (continued)

	Research paper (chaps. 10 and 11)	20-min. oral presentation (chap. 12)	4' by 8' poster presentation (chap. 13)
			& policy implications, to be presented as appropriate in response to questions from individual viewers. Include "Vanna White" notes as needed
			A few questions to ask viewers to solicit reactions to your findings, ideas for additional questions, or names of others working on the topic
Handouts	Not applicable	Handout of slides, copied several to a page For longer seminars: abstract and detailed tables	Project title, contact information, abstract & handout of slides, copied several to a page
Interaction with audience	Occasional e-mail, phone, or in-person queries	Questions and discussion from the audience • Usually quite limited in a multi-paper conference session • Can be extensive during a longer individual seminar • Questions from individuals after the session; tailor responses to each person	One-on-one discussion with viewers, tailored in response to questions and responses from each viewer; see "Spoken materials" above Potentially extensive, depending on interests of viewers

Table B.2. Comparison of research papers, oral presentations, and posters—contents

	Research paper[a] (chaps. 10 and 11)	20-min. oral presentation[ab] (chap. 12)	4' by 8' poster presentation (chap. 13)
Abstract	Yes	No	Yes, or revise into "What We Learned"
Introduction	Several pages of background on issue and its importance, ending with statement of research question or hypotheses.	1–2 slides of background on issue and its importance. One slide stating research question or hypotheses.	1–2 pages of background on issue and its importance, ending with statement of research question or hypotheses
Literature review	Detailed review and summary of previous studies on similar topics and methods.	1–2 slides of few key studies only, as tabular summary (e.g., figure 12.4) or bulleted text. More detail for longer seminars.	1 page, focusing on main points from few key articles and identifying gaps in the literature
Data and methods	Comprehensive, detailed information on data sources, study design, variables, and statistical methods. May include equations.	3–4 slides presenting essential information on data sources, study design, variables, and statistical methods.	2–3 pages presenting essential information on data sources, study design, variables, and statistical methods

(continued)

Table B.2. (*continued*)

	Research paper[a] (chaps. 10 and 11)	20-min. oral presentation[ab] (chap. 12)	4′ by 8′ poster presentation (chap. 13)
Results	Detailed statistical tables and charts accompanied by prose descriptions written in paragraph form.	4–5 slides with simplified tables and charts, accompanied by bulleted text annotations or described in speaker's notes. One major result or set of related results per slide, reflected in slide title.	4–5 pages of simplified tables and charts accompanied by bulleted text annotations One major result or set of related results per page, reflected in page title
Conclusions	Several pages relating findings to research question and to related studies, discussing study strengths and limitations, and describing research and policy implications.	One slide each of bulleted text on • Summary of key findings • Policy or practice implications • Directions for future research • Strengths and limitations	2–3 pages summarizing key findings, discussing study strengths and limitations, and describing implications for research, policy, and practice

[a] Assumes a professional research audience. Read instructions for authors for your intended journal to obtain guidelines about overall length and content of specific sections. For lay audiences, reduce emphasis on data, methods, statistical results, and research implications; increase emphasis on purpose, findings, and policy implications; see chapter 13.

[b] For a longer presentation or seminar, increase sections proportionately.

NOTES

Chapter 2

1. See Best 2001.
2. See chapter 3 for further discussion of various dimensions of "significance" that come into play when assessing quantitative relations.
3. Another aspect of association—statistical significance—is covered in chapter 3.

Chapter 3

1. The fifth criterion—specificity—is most applicable to the study of infectious disease. It concerns the extent to which a particular exposure (e.g., the measles virus) produces one specific disease (measles).
2. The value 1.96 is the critical value for $p < 0.05$ with a two-tailed test and a large sample size. For most purposes, twice the standard error gives a good approximation (and is much easier to calculate).
3. Other width confidence intervals can also be calculated. For example, a 99% CI (equivalent to testing whether $p < 0.01$) is calculated as the point estimate ± 2.58 times the standard deviation.

Chapter 4

1. Temperature in degrees Kelvin has a meaningful absolute zero value and can be treated as a ratio variable, but is rarely used by anyone other than physical scientists in technical documents.
2. When categorical variables are entered into an electronic database, each group is often assigned a numeric code as an abbreviation. Do not treat those values as if they had real numeric meaning. It makes no sense to calculate "average gender" by computing the mean of a variable coded 1 for male and 2 for female. For categorical variables, the appropriate measure of central tendency is the mode, not the mean.
3. "Index" and "scale" are examples of terms that are used differently in different fields (Babbie 2006; Treiman 2009; Chambliss and Schutt 2012). See a textbook in your field for clarification about which term to use for your particular application and audience.
4. To generalize, "____ % of [concept in the denominator] is [concept in the numerator]."
5. Sometimes the geometric mean is used instead of the arithmetic mean. It is computed as the nth root of the product of all values in the sample, where n is the number of values in the calculation. If you use the geometric mean, explicitly name it to avoid confusion.
6. In the phrase "significant digits," the term "significant" has a different mean-

ing from the statistical interpretation discussed in chapter 3. Here, it refers to precision of measurement and how that affects the appropriate number of digits in measured values (raw data) and calculations.

Chapter 5

1. For authors working with multivariate analyses, some additional types of comparison are needed. See Miller (2013a) for a detailed discussion of these calculations and their interpretation.

2. In the first edition of *The Chicago Guide to Writing about Numbers* (2004), I termed this the "absolute difference." An astute reviewer pointed out that "absolute value" connotes the distance from 0, regardless of sign (Kornegay 1999). For instance, −2 and +2 both have an absolute value (difference from 0) of 2 units. In order to avoid confusing this meaning of "absolute difference" with my intended use, I henceforth refer to the results of subtraction as simply the "difference."

3. Changing the reference group in a two-group ratio calculation merely involves "flipping over" the ratio to calculate its reciprocal: a ratio of 1.56 southerners per midwesterner is equivalent to 0.64 midwesterners per southerner.

4. With negative growth rates (yes, they are called that, not "shrinkage rates"!), the base population, or principal, becomes successively smaller across time.

5. For annual compounding, calculate the annual interest rate (r) using the formula $\log(1 + r) = \log(P_2/P_1)/n$, where P_1 and P_2 are the populations at times 1 and 2 respectively, n is the number of years between P_1 and P_2, and "log" indicates base 10 logarithms. For continuous compounding, use $r = \ln(P_2/P_1)/n$, where "ln" indicates natural logarithms (Barclay 1958).

6. A relative risk is a ratio of two risks or rates. For instance, the relative risk of cancer for people exposed to a specific toxin compared to those who have not been exposed would divide the rate of cancer in the exposed group by the rate of cancer in the unexposed group.

Chapter 6

1. Breakdowns of non-Hispanic blacks and non-Hispanics of other races could likewise be indented under the respective headings. Those data were not available in the current source (US Census Bureau 1998).

2. For example, as shown in table 6.6, there were 43,287 thousand people aged 65 in the United States in 2012, of whom 9.1% were poor. Hence 43,287,000 × 0.091 = 3,939 thousand elderly persons were poor in 2012.

3. Place the category that combines "all other" values at the bottom of the table body, before the "total" row. Frequently, it is not ranked with specifically named categories, as it combines several values.

4. However, if a response is missing for a substantial share of cases, show the

distribution of both "yes" and "no," as well as "don't know" or other missing values.

Chapter 7

1. A pie chart can present two categorical variables simultaneously by cross-tabulating them. E.g., a single pie showing gender and age distribution might have slices for males under 20, males 20 and older, females under 20, and females 20 and older—four mutually exclusive categories. Univariate slices (e.g., one slice for <20 and another for males) cannot be shown in the same pie chart because some people are both <20 and male.
2. Some spreadsheet programs have an error bar feature that can also be used to create such a chart.
3. Ninety-five percent confidence intervals are the standard, corresponding to a p-value of 0.05. If you use a different confidence level such as 90% or 99%, include that information in the chart title.

Chapter 10

1. In addition to the deaths that occurred within a year of each survey date, other cases were lost from the sample due to other forms of attrition (e.g., moving away or refusing to participate) or death in intervening years (Idler et al. 2001).

Chapter 12

1. For audiences interested in standard errors or other ways of presenting statistical significance that allow them to make their own comparisons (see Miller 2013a, chapter 11), hand out the full statistical table at the end of the speech for readers to peruse on their own time. If you distribute it during the talk, they'll pay attention to it, not you, and may ask distracting questions about specific numbers unrelated to your main points.
2. The "Vanna White" moniker is in honor of the longtime hostess of the TV game show *Wheel of Fortune*, who gestures at the display to identify each item or feature as it is introduced.

Chapter 13

1. With the increasing availability of printers that can produce large (3' by 6' or larger) single-sheet posters, talking about "pages" can be somewhat confusing. In the guidelines about posters I use the word "page" to refer to the component sections of a poster, each of which will often be approximately the size of a standard 8.5" by 11" printed page.

REFERENCE LIST

Abramson, J. H. 1994. *Making Sense of Data: A Self-Instruction Manual on the Interpretation of Epidemiologic Data*. 2nd ed. New York: Oxford University Press.

Agresti, Alan, and Barbara Finlay. 1997. *Statistical Methods for the Social Sciences*. 3rd ed. Upper Saddle River, NJ: Prentice Hall.

Allison, Paul D. 1999. *Multiple Regression: A Primer*. Thousand Oaks: Sage Publications.

Alred, Gerald J., Charles T. Brusaw, and Walter E. Oliu. 2000. *Handbook of Technical Writing*. 6th ed. New York: St. Martin's Press.

American Heart Association Statistics Committee and Stroke Statistics Subcommittee. 2014. Heart Disease and Stroke Statistics—2014 Update: A Report from the American Heart Association. *Circulation* 129 (2014): e28–e292.

American Medical Association. 2007. *AMA Manual of Style: A Guide for Authors and Editors*. 10th ed. Oxford: Oxford University Press.

American Psychological Association. 2009. *Publication Manual of the American Psychological Association*. 6th ed. Washington DC: American Psychological Association.

Anderson, Robert N., and Harry M. Rosenberg. 1998. "Age Standardization of Death Rates: Implementation of the Year 2000 Standard." *National Vital Statistics Report* 47 (3). Hyattsville, MD: National Center for Health Statistics.

Babbie, Earl R. 2006. *The Practice of Social Research*. 11th ed. Florence, KY: Wadsworth Publishing.

Barclay, George W. 1958. *Techniques of Population Analysis*. New York: John Wiley and Sons.

Benson, Veronica, and Marie A. Marano. 1998. "Current Estimates from the National Health Interview Survey, 1995." *Vital and Health Statistics* 10 (199). Hyattsville, MD: National Center for Health Statistics.

Best, Joel. 2001. *Damned Lies and Statistics*. Berkeley: University of California Press.

Briscoe, Mary Helen. 1996. *Preparing Scientific Illustrations: A Guide to Better Posters, Presentations, and Publications*. 2nd ed. New York: Springer-Verlag.

Centers for Disease Control and Prevention (CDC). 1999. "Notice to Readers: New Population Standard for Age-Adjusting Death Rates." *Morbidity and Mortality Weekly Review* 48 (6): 126–27. Available at http://www.cdc.gov /mmwr/preview/mmwrhtml/00056512.htm.

———. 2002. "Compressed Mortality File: U.S. 1999."

Centers for Medicare and Medicaid Services, Office of the Actuary, National Health Statistics Group, Department of Health and Human Services. 2013.

"Historical National Health Expenditure Data, Table 1. National Health Expenditures; Aggregate and Per Capita Amounts, Annual Percent Change and Percent Distribution: Selected Calendar Years 1960–2012." Available at http://www.cms.gov/Research-Statistics-Data-and-Systems/Statistics -Trends-and-Reports/NationalHealthExpendData/Downloads/tables.pdf. Accessed January 2014.

Chambliss, Daniel F., and Russell K. Schutt. 2012. *Making Sense of the Social World: Methods of Investigation.* 4th ed. Thousand Oaks, CA: Sage Publications.

Citro, Constance F., Robert T. Michael, and Nancy Maritano, eds. 1996. *Measuring Poverty: A New Approach.* Table 5–8. Washington, DC: National Academy Press.

Conley, Dalton, and Neil G. Bennett. 2000. "Race and the Inheritance of Low Birth Weight." *Social Biology* 47 (1–2): 77–93.

Cornoni-Huntley, J., D. B. Brock, A. M. Ostfeld, J. O. Taylor, and R. B. Wallace, eds. 1986. *Established Populations for Epidemiologic Studies of the Elderly: Resource Data Book.* NIH Publ. no. 86–2443. Bethesda, MD: National Institute on Aging.

Davis, James A. 1985. *The Logic of Causal Order.* Quantitative Applications in the Social Sciences 55. Thousand Oaks, CA: Sage Publications.

Davis, James A., Tom W. Smith, and Peter V. Marsden. 2003. *General Social Surveys, 1972–2000: Computer Codebook.* 2nd ICPSR version. Chicago: National Opinion Research Center.

Davis, Martha. 1997. *Scientific Papers and Presentations.* New York: Academic Press.

DeNavas-Walt, Carmen, Bernadette D. Proctor, and Jessica C. Smith. 2013. US Census Bureau, Income, Poverty, and Health Insurance Coverage in the United States: 2012. *Current Population Reports,* P60–245. Washington, DC: US Government Printing Office.

Dewdney, A. K. 1996. *200% of Nothing: An Eye-Opening Tour through the Twists and Turns of Math Abuse and Innumeracy.* New York: John Wiley and Sons.

DiFranza, Joseph R., and the Staff of the Advocacy Institute, with assistance from the Center for Strategic Communications. 1996. *A Researcher's Guide to Effective Dissemination of Policy-Related Research.* Princeton, NJ: Robert Wood Johnson Foundation.

Duly, Abby. 2003. "Consumer Spending for Necessities." *Monthly Labor Review* 126 (5): 3. Available at http://stats.bls.gov/opub/mlr/2003/05/art1full.pdf. Accessed August 2004.

Eaton, Leslie. 2002. "Job Data May Show a False Positive." *New York Times,* February 10, Job Market section.

Fink, Arlene. 1995. *How to Report on Surveys*. Thousand Oaks, CA: Sage Publications.

Franzini, L., J. C. Ribble, and A. M. Keddie. 2001. "Understanding the Hispanic Paradox." *Ethnicity and Disease* 11 (3): 496–518.

Githens, P. B., C. A. Glass, F. A. Sloan, and S. S. Entman. 1993. "Maternal Recall and Medical Records: An Examination of Events during Pregnancy, Childbirth, and Early Infancy." *Birth* 20 (3): 136–41.

Glanz, James, and Eric Lipton. 2002. "The Height of Ambition." *New York Times*, September 8.

Gold, Marthe R., Joanna E. Siegel, Louise B. Russell, and Milton C. Weinstein, eds. 1996. *Cost-Effectiveness in Health and Medicine*. New York: Oxford University Press.

Hailman, Jack P., and Karen B. Strier. 1997. *Planning, Proposing, and Presenting Science Effectively: A Guide for Graduate Students and Researchers in the Behavioral Sciences and Biology*. New York: Cambridge University Press.

Hoaglin, David C., Frederick Mosteller, and John W. Tukey. 2000. *Understanding Robust and Exploratory Data Analysis*. New York: John Wiley and Sons.

Idler, Ellen L., Stanislav V. Kasl, and Judith C. Hays. 2001. "Patterns of Religious Practice and Belief in the Last Year of Life." *Journal of Gerontology* 56B (6): S326–S334.

Institute of Medicine, Committee to Study the Prevention of Low Birthweight. 1985. *Preventing Low Birthweight*. Washington, DC: National Academy Press.

International Institute for Democracy and Electoral Assistance. 1999. "Voter Turnout from 1945 to 1998: A Global Participation Report." Available at http://www.int-idea.se/Voter_turnout/northamerica/usa.html. Accessed January 2003.

Johnson, Kirk. 2004. "Colorado Takes Steps to Polish Thin and Fit Image." *New York Times*, February 1, Front section.

Kolata, Gina. 2003. "Hormone Studies: What Went Wrong?" *New York Times*, April 22, Science section.

Kornegay, Chris. 1999. *Math Dictionary with Solutions: A Math Review*. 2nd ed. Thousand Oaks, CA: Sage Publications.

Kosslyn, Stephen M. 2007. *Clear and to the Point: 8 Psychological Principles for Compelling PowerPoint Presentations*. New York: Oxford University Press.

Kraemer, Helena C. 1987. *How Many Subjects? Statistical Power Analysis in Research*. Thousand Oaks, CA: Sage Publications.

Kuczmarski, Robert J., et al. 2000. "CDC Growth Charts: United States." *Advance Data from Vital and Health Statistics*. No. 314. Hyattsville, MD: National Center for Health Statistics.

Lanham, Richard. 2000. *Revising Prose*. 4th ed. New York: Longham.

Leonhardt, David. 2003. "For Aging Runners, a Formula Makes Time Stand Still." *New York Times*, October 28.

Levy, Frank. 1987. *Dollars and Dreams: The Changing American Income Distribution*. New York: Russell Sage Foundation.

Levy, Helen, and Thomas DeLeire. 2002. "What Do People Buy When They Don't Buy Health Insurance?" Northwestern University/University of Chicago Joint Center for Poverty Research Working Paper 296 06–26-2002.

Lewit, Eugene, L. Baker, Hope Corman, and P. Shiono. 1995. "The Direct Cost of Low Birth Weight." *Future of Children* 5 (1).

Lilienfeld, David E., and Paul D. Stolley. 1994. *Foundations of Epidemiology*. 3rd ed. New York: Oxford University Press.

Lipton, Eric, and James Glanz. 2002. "First Tower to Fall Was Hit at Higher Speed, Study Finds." *New York Times*, February 23.

Logan, Ralph H. 1995. "Significant Digits." Available at http://members.aol.com /profchm/sig_fig.html. Accessed January 2003.

Ma, Xin, and J. Douglas Willms. 1999. "Dropping Out of Advanced Mathematics: How Much Do Students and Schools Contribute to the Problem?" *Educational Evaluation and Policy Analysis* 21: 365–83.

Maciejewski, Matthew L., Paula Diehr, Maureen A. Smith, and Paul Hebert. 2002. "Common Methodological Terms in Health Services Research and Their Symptoms." *Medical Care* 40: 477–84.

Mack, Tim. n.d. "Plagues since the Roman Empire." Available at http://www .ento.vt.edu/IHS/plagueHistory.html#justinian. Accessed January 2002.

Martin, Joyce A., Brady E. Hamilton, Stephanie J. Ventura, Fay Menacker, and Melissa M. Park. 2002. "Births: Final Data for 2000." *National Vital Statistics Reports* 50 (5).

Mathews, T. J., and Brady E. Hamilton. 2002. "Mean Age of Mother, 1970–2000." *National Vital Statistics Report* 51 (1). Hyattsville, MD: National Center for Health Statistics.

Mathews, T. J., Marian F. MacDorman, and Fay Menacker. 2002. "Infant Mortality Statistics from the 1999 Period Linked Birth/Infant Death Data Set." *National Vital Statistics Reports* 50 (4). Hyattsville, MD: National Center for Health Statistics.

McDonough, John. 2000. *Experiencing Politics: A Legislator's Stories of Government and Health Care*. Berkeley: University of California Press.

Miller, Jane E. 2000a. "Differences in AIDS Knowledge among Spanish and English Speakers by Socioeconomic Status and Ability to Speak English." *Journal of Urban Health* 77 (3): 415–24.

———. 2000b. "Income, Race/Ethnicity, and Early Childhood Asthma Prevalence and Health Care Utilization." *American Journal of Public Health* 90 (3): 428–30.

———. 2006. "How to Communicate Statistical Findings: An Expository Writing Approach." *Chance* 19 (4): 43–49.

———. 2007a. "Organizing Data in Tables and Charts: Different Criteria for Different Tasks." *Teaching Statistics* 29 (3): 98–101.

———. 2007b. "Preparing and Presenting Effective Research Posters." *Health Services Research* 42 (No. 1, Part I): 311–28.

———. 2010. "Quantitative Literacy across the Curriculum: Integrating Skills from English Composition, Mathematics, and the Substantive Disciplines." *Educational Forum* 74 (4): 334–46.

———. 2013a. *The Chicago Guide to Writing about Multivariate Analysis.* 2nd ed. Chicago: University of Chicago Press.

———. 2013b. "Getting to Know Your Variables: The Foundation for a Good Working Relationship with Your Data." In *2013 Proceedings of the American Statistical Association, Statistical Computing Section,* pp. 2016–27. Alexandria, VA: American Statistical Association.

———. 2014. "Planning How to Create the Variables You Need from the Variables You Have." Working paper, Rutgers University.

Miller, Jane E., and Yana V. Rodgers. 2008. "Economic Importance and Statistical Significance: Guidelines for Communicating Empirical Research." *Feminist Economics* 14 (2): 117–49.

Monmonier, Mark. 1993. *Mapping It Out: Expository Cartography for the Humanities and Social Sciences.* Chicago: University of Chicago Press.

Montgomery, Scott L. 2003. *The Chicago Guide to Communicating Science.* Chicago: University of Chicago Press.

Mooney, Christopher Z., and Mei Hsien Lee. 1995. "Legislating Morality in the American States: The Case of Abortion Regulation Reform." *American Journal of Political Science* 39: 599–627.

Moore, David S. 1997. *Statistics: Concepts and Controversies.* 4th ed. New York: W. H. Freeman and Co.

Morgan, Susan E., Tom Reichert, and Tyler R. Harrison. 2002. *From Numbers to Words: Reporting Statistical Results for the Social Sciences.* Boston: Allyn and Bacon.

Morton, Richard F., J. Richard Hebel, and Robert J. McCarter. 2001. *A Study Guide to Epidemiology and Biostatistics.* 5th ed. Gaithersburg, MD: Aspen Publishers.

Murphy, Kevin R. 2010. "Power Analysis." In *The Reviewer's Guide to Quantitative Methods in the Social Sciences,* ed. Gregory R. Hancock and Ralph O. Mueller. Florence, KY: Routledge.

Murphy, Sherry L., Jiaquan Xu, and Kenneth D. Kochanek. 2012. "Deaths: Preliminary Data for 2010." *National Vital Statistics Reports* 60 (4). Hyattsville, MD: National Center for Health Statistics. Available at http://www.cdc.gov/nchs/data/nvsr/nvsr60/nvsr60_04.pdf.

Musso, Juliet, Robert Biller, and Robert Myrtle. 2000. "Tradecraft: Professional Writing as Problem Solving." *Journal of Policy Analysis and Management* 19 (4): 635–46.

National Center for Health Statistics (NCHS). 1994. "Plan and Operation of the Third National Health and Nutrition Examination Survey, 1988–1994." *Vital and Health Statistics* 1 (32).

———. 1995. "International Classification of Diseases, Ninth Revision, Clinical Modification." 5th ed. NCHS CD-ROM no. 1. DHHS pub. no. (PHS) 95–1260.

———. 2002. "Policy on Micro-Data Dissemination." Available at http://www .cdc.gov/nchs/data/NCHS%20Micro-Data%20Release%20Policy%204–02A .pdf. Accessed January 2003.

———. 2013. *Health, United States, 2012: With Special Feature on Emergency Care.* Hyattsville, MD: National Center for Health Statistics. Available at http:// www.cdc.gov/nchs/data/hus/hus12.pdfAccessed January 2014.

National Institutes for Health, Office of Human Subjects Research. 2002. *Regulations and Ethics.* Available at http://206.102.88.10/ohsrsite /guidelines/guidelines.html. Accessed January 2003.

National Insurance Crime Bureau (NICB). 2011. *Hot Wheels.* Available at https://www.nicb.org/. Accessed June 2012.

———. 2012. *Hot Wheels Classics: Chevrolet Corvette; A Truly Hot Car—More than One in 10 Stolen over Past 30 Years.* Available at https://www.nicb.org /newsroom/news-releases/chevrolet-corvette-thefts. Accessed June 2012.

National Opinion Research Center. 2000. "2000 General Social Survey Questionnaire, version 1." Chicago: NORC, University of Chicago. Available at http://publicdata.norc.org/GSS/DOCUMENTS/QUEX/2000/2000%20 GSS%20v1.pdf. Accessed January 2014.

Navarro, Mireya. 2003. "Going beyond Black and White, Hispanics in Census Pick 'Other.'" *New York Times,* November 9.

Nelson, David E., Ross C. Brownson, Patrick L. Remington, and Claudia Parvanta, eds. 2002. *Communicating Public Health Information Effectively: A Guide for Practitioners.* Washington, DC: American Public Health Association.

Nicol, Adelheid A. M., and Penny M. Pexman. 1999. *Presenting Your Findings: A Practical Guide for Creating Tables.* Washington, DC: American Psychological Association.

NIST (National Institute of Standards and Technology). 2000. "Uncertainty of Measurement Results." In *The NIST Reference on Constants, Units, and Uncertainty.* Available at http://physics.nist.gov/cuu/Uncertainty/index .html. Accessed January 2004.

Olson, J. E., X. O. Shu, J. A. Ross, T. Pendergrass, and L. L. Robison. 1997. "Medical Record Validation of Maternally Reported Birth Characteristics and

Pregnancy-Related Events: A Report from the Children's Cancer Group."
American Journal of Epidemiology 145 (1): 58–67.

Omran, Abdel R. 1971. "The Epidemiologic Transition: A Theory of the
Epidemiology of Population Change." *Milbank Memorial Fund Quarterly* 49
(4): 509–38.

Organisation for Economic Co-operation and Development (OECD). 2013.
Health at a Glance 2013: OECD Indicators. Paris: OECD Publishing. Available
at http://dx.doi.org/10.1787/health_glance-2013-en. Accessed January
2014.

Palloni, Alberto, and J. D. Morenoff. 2001. "Interpreting the Paradoxical in the
Hispanic Paradox: Demographic and Epidemiologic Approaches." *Annals of
the New York Academy of Sciences* 954: 140–74.

Pan, M. Ling. 2008. *Preparing Literature Reviews: Qualitative and Quantitative
Approaches.* 3rd ed. Glendale, CA: Pyrczak Publishing.

Paneth, Nigel. 1995. "The Problem of Low Birth Weight." *Future of Children* 5 (1).

Paulos, John A. 2001. *Innumeracy: Mathematical Illiteracy and Its Consequences.*
New York: Farrar, Straus, and Giroux.

Peters, Kimberley D., Kenneth D. Kochanek, and Sherry L. Murphy. 1998.
"Deaths: Final Data for 1996." *National Vital Statisitcs Report* 47 (9).
Available at http://www.cdc.gov/nchs/data/nvsr/nvsr47/nvs47_09.pdf.
Accessed August 2014.

Phillips, Anna M. 2012. "Number of Elementary School Students in Large
Classes Has Tripled, a Councilman Says." *New York Times.* March 27,
Section A, p. 24.

Pollack, Andrew. 1999. "Missing What Didn't Add Up, NASA Subtracted an
Orbiter." *New York Times,* October 1.

Population Reference Bureau. 1999. "World Population: More than Just
Numbers." Washington, DC: Population Reference Bureau.

Pottick, Kathleen J., and Lynn A. Warner. 2002. "More than 115,000
Disadvantaged Preschoolers Receive Mental Health Services." *Update: Latest
Findings in Children's Mental Health* 1 (2). New Brunswick, NJ: Institute for
Health, Health Care Policy, and Aging Research. Available at http://www
.ihhcpar.rutgers.edu/downloads/fa112002.pdf. Accessed May 2004.

Preston, Samuel H. 1976. *Mortality Patterns in National Populations.* New York:
Academic Press.

Proctor, Bernadette D., and Joseph Dalaker. 2002. "Poverty in the U.S., 2001."
Current Population Reports. P60–219. Washington, DC: US Government
Printing Office.

Puffer, Ruth R., and Carlos V. Serrano. 1973. *Patterns of Mortality in Childhood.*
Washington, DC: World Health Organization.

Pyrczak, Fred, and Randall R. Bruce. 2011. *Writing Empirical Research Reports*. 7th ed. Los Angeles: Pyrczak Publishing.

Quality Resource Systems, Inc. 2001. *Area Resource File*. Available at http://www.arfsys.com/.

Radloff, L. S., and B. Z. Locke. 1986. "The Community Mental Health Assessment Survey and the CES-D Scale." In *Community Surveys of Psychiatric Disorders*, edited by M. M. Weissman, J. K. Myers, and C. E. Ross. New Brunswick, NJ: Rutgers University Press.

Schwartz, John. 2003. "The Level of Discourse Continues to Slide." *New York Times*, Week in Review, September 28.

Shah, B. V., B. G. Barnwell, and G. S. Bieler. 1996. *SUDAAN User's Manual, Release 7.0*. Research Triangle Park, NC: Research Triangle Institute.

Slocum, Terry A. 1998. *Thematic Cartography and Visualization*. Upper Saddle River, NJ: Prentice Hall.

Smith, Philip J., Diane Simpson, Michael P. Battaglia, and Abt Associates Inc. 2000. *Split Sampling Design for Topical Modules in the National Immunization Survey*. Atlanta, GA: Centers for Disease Control and Prevention. Available at http://www.cdc.gov/nchs/data/nis/miscellaneous/smith2000d.pdf. Accessed October 2012.

Snow, John. 1936. *Snow on Cholera*. New York: Commonwealth Fund.

Snowdon, David. 2001. *Aging with Grace: What the Nun Study Teaches Us about Leading Longer, Healthier, and More Meaningful Lives*. New York: Bantam Books.

Strunk, William, Jr., and E. B. White. 1999. *The Elements of Style*. 4th ed. Boston: Allyn and Bacon.

Sorian, Richard, and Terry Baugh. 2002. "Power of Information: Closing the Gap between Research and Policy." *Health Affairs* 21 (2): 264–73.

Steen, Lynn A., ed. 2001. *Mathematics and Democracy: The Case for Quantitative Literacy*. Princeton, NJ: Woodrow Wilson National Foundation.

———. 2004. *Achieving Quantitative Literacy: An Urgent Challenge for Higher Education*. Mathematical Association of America.

Tavernise, Sabrina. 2014. "Young Using E-Cigarettes Smoke Too, Study Finds." *New York Times*, March 6, p. A17.

Thompson, Bruce. 2004. "The 'Significance' Crisis in Psychology and Education." *Journal of Socio-Economics* 33 (5): 607–13.

Treiman, Donald J. 2009. *Quantitative Data Analysis: Doing Social Research to Test Ideas*. San Francisco: Jossey-Bass.

Tufte, Edward R. 1990. *Envisioning Information*. Cheshire, CT: Graphics Press.

———. 1997. *Visual Explanations: Images and Quantities, Evidence and Narrative*. Cheshire, CT: Graphics Press.

———. 2001. *The Visual Display of Quantitative Information.* 2nd ed. Cheshire, CT: Graphics Press.

———. 2003. *The Cognitive Style of PowerPoint.* Cheshire, CT: Graphics Press.

Tukey, John W. 1977. *Exploratory Data Analysis.* New York: Addison-Wesley Publishing Co.

University of Chicago Press. 2010. *The Chicago Manual of Style: The Essential Guide for Writers, Editors, and Publishers.* 16th ed. Chicago: University of Chicago Press.

US Bureau of Labor Statistics. 2007. "The Consumer Price Index." In *BLS Handbook of Methods.* Available at http://www.bls.gov/opub/hom/. Accessed June 2012.

———. 2012. "Regional and State Employment and Unemployment—February 2012." Available at http://www.bls.gov/news.release/laus.nr0.htm. Accessed April 2012.

US Census Bureau. 1998. "Households by Type and Selected Characteristics: 1998." Available at http://www.census.gov/population/socdemo/hh-fam/98ppla.txt.

———. 2001a. "Table 940. Median Sales Price of New Privately Owned One-Family Houses Sold by Region, 1980–2000." *Statistical Abstract of the United States, 2001.* Available at http://www.census.gov/prod /2002pubs/01statab/construct.pdf.

———. 2001b. "Money Income in the United States: 2000." *Current Population Reports.* P60–213. Washington, DC: US Government Printing Office.

———. 2002a. "Poverty 2001." Available at http://www.census.gov/hhes /poverty/threshld/thresh01.html. Accessed October 2002.

———. 2002b. "No. 310. Drug Use by Arrestees in Major United States Cities by Type of Drug and Sex, 1999." *Statistical Abstract of the United States, 2001.* Available at http://www.census.gov/prod/2002pubs/01statab/stat-ab01 .html. Accessed January 2004.

———. 2002c. "Table 611. Average Annual Pay, by State, U.S. 1999 and 2000." *Statistical Abstract of the United States, 2002.*

———. 2002d. "2000 Detailed Tables: Sex by Age, U.S. Summary File SF1." Available at http://factfinder.census.gov/. Accessed September 2002.

———. 2010. "Table 4. Resident Population of the 50 States, the District of Columbia, and Puerto Rico: 2010 Census and Census 2000." Available at http://2010.census.gov/news/xls/apport2010_table4.xls.

———. 2012. "Table 587. Civilian Labor Force and Participation Rates with Projections: 1980 to 2018." *Statistical Abstract of the United States: 2012.* 131st ed. Washington, DC, 2011. Available at http://www.census.gov /compendia/statab/2012/tables/12s0587.pdf.

US Census Bureau. 2014. US Census Bureau: State and Country QuickFacts. Available at http://quickfacts.census.gov/qfd/states/00000.html.

US Department of Labor, Bureau of Labor Statistics. 2004. *Consumer Expenditures in 2002.* Report 974. Available at http://stats.bls.gov/cex /csxann02.pdf. Accessed August 2004.

US DHHS (Department of Health and Human Services). 1991. *Longitudinal Follow-Up to the 1988 National Maternal and Infant Health Survey.* Hyattsville, MD: National Center for Health Statistics.

———. 1997. *National Health and Nutrition Examination Survey, III, 1988–1994.* CD-ROM Series 11, no. 1. Hyattsville, MD: National Center for Health Statistics, Centers for Disease Control and Prevention.

———, Office of Disease Prevention and Health Promotion. 2001. *Healthy People 2010.* US GPO Stock no. 017–001-00547–9. Washington, DC: US Government Printing Office. Available at http://www.health.gov/healthypeople/. Accessed September 2002.

———, Centers for Disease Control and Prevention. 2002. *Z-score Data Files.* Hyattsville, MD: National Center for Health Statistics, Division of Data Services. Available at http://www.cdc.gov/nchs/about/major/nhanes /growthcharts/zscore/zscore.htm. Accessed January 2003.

———, Office of Human Service Policy. 2013. *Indicators of Welfare Dependence: Twelfth Report to Congress.* Washington, DC. Available at http://aspe.hhs .gov/hsp/13/Indicators/rpt.pdf Accessed January 2014.

US Environmental Protection Agency. 2002. *Cost of Illness Handbook.* Available at http://www.epa.gov/oppt/coi/. Accessed September 2002.

US National Energy Information Center. 2003. "Crude Oil Production. Table 2.2. World Crude Oil Production, 1980–2001." Available at http://www.eia.doe .gov/pub/international/iealf/table22.xls. Accessed January 2004.

US Office of Management and Budget. 2002. *Budget of the United States, Fiscal Year 2003, Historical Tables.* Table 3.1. Available at http://w3.access.gpo.gov /usbudget/fy2003/pdf/hist.pdf. Accessed June 2002.

US Social Security Administration. 2012. *Income of the Aged Chartbook, 2010.* Baltimore, MD. Available at http://www.ssa.gov/policy/docs/chartbooks /income_aged/2010/iac10.pdf. Accessed June 2012.

Utts, Jessica M. 1999. *Seeing through Statistics.* 2nd ed. New York: Duxbury Press.

Ventura, Stephanie J., Joyce A. Martin, Sally C. Curtin, and T. J. Mathews. 1999. "Births: Final Data for 1997." *National Vital Statistics Report* 47 (18). Hyattsville, MD: National Center for Health Statistics.

Warner, Lynn A., and Kathleen J. Pottick. 2004. "More than 380,000 Children Diagnosed with Multiple Mental Health Problems." In *Latest Findings in Children's Mental Health,* 3 (1). New Brunswick, NJ: Institute for Health, Health Care Policy, and Aging Research. Available at http://www.ihhcpar .rutgers.edu/downloads/winter2004.pdf. Accessed January 2014.

Westat. 1994. "National Health and Nutrition Examination Survey III, Section 4.1: Accounting for Item Non-Response Bias." In *NHANES III Reference Manuals and Reports*. Washington, DC: US Department of Health and Human Services, Public Health Service, Centers for Disease Control and Prevention, National Center for Health Statistics.

———. 1996. "National Health and Nutrition Examination Survey III, Weighting and Estimation Methodology." In *NHANES III Reference Manuals and Reports*. Hyattsville, MD: US Department of Health and Human Services, Public Health Service, Centers for Disease Control and Prevention, National Center for Health Statistics.

Weston, Harley. n.d. "95% Confidence Intervals." MathCentral. Available at http://mathcentral.uregina.ca/QQ/database/QQ.09.07/h/fara1.html. Accessed January 2014.

Wilkinson, Leland, and Task Force on Statistical Inference, APA Board of Scientific Affairs. 1999. "Statistical Methods in Psychology Journals: Guidelines and Explanations." *American Psychologist* 54 (8): 594–604.

Willingham, Daniel T. 2009. "Why Is It So Hard for Students to Understand Abstract Ideas?" In *Why Don't Students Like School? A Cognitive Scientist Answers Questions about How the Mind Works and What It Means for the Classroom*. San Francisco: Jossey-Bass.

Wolke, Robert L. 2002. *What Einstein Told His Cook: Kitchen Science Explained*. New York: W. W. Norton.

World Bank. 2012. "Indicators." Available at http://data.worldbank.org/indicator. Accessed April 2012.

World Health Organization, Expert Committee on Maternal and Child Health. 1950. Public Health Aspect of Low Birthweight. WHO Technical Report Series, no. 27. Geneva: WHO.

———. 1995. *Physical Status: Use and Interpretation of Anthropometry*. WHO Technical Report Series, no. 854. Geneva: WHO.

———. 2002. "Quantifying Selected Major Risks to Health." In *World Health Report 2002: Reducing Risks, Promoting Healthy Life*. Geneva: WHO.

www.fueleconomy.gov. 2012. "Find and Compare Cars." Available at http://www.fueleconomy.gov/feg/findacar.shtml. Accessed April 2012.

Young, L. R., and M. Nestle. 2002. "The Contribution of Expanding Portion Sizes to the US Obesity Epidemic." *American Journal of Public Health* 92 (2): 246–49.

Zambrana, R. E., C. Dunkel-Schetter, N. L. Collins, and S. C. Scrimshaw. 1999. "Mediators of Ethnic-Associated Differences in Infant Birth Weight." *Journal of Urban Health* 76 (1): 102–16.

Zelazny, Gene. 2001. *Say It with Charts: The Executive's Guide to Visual Communication*. 4th ed. New York: McGraw-Hill.

Ziliak, Stephen T., and Deirdre N. McCloskey. 2004. "Size Matters: The Standard Error of Regressions in the American Economic Review." *Journal of Socio-Economics* 33 (5): 527–46.

———. 2008. *The Cult of Statistical Significance: How the Standard Error Costs Us Jobs, Justice, and Lives.* Ann Arbor: University of Michigan Press.

Zinsser, William. 1998. *On Writing Well: The Classic Guide to Writing Nonfiction.* 6th ed. New York: HarperCollins Publishers.

INDEX

A page number followed by *f* refers to a figure, and a page number followed by *t* indicates a table.

appendices, 21, 127, 129, 144, 149
application (intended use) of
    example, 198. *See also* examples
applied audiences
  assessing specific audience, 36,
    317, 328
  charts for, 324. *See also* charts
  checklist for, 346–47
  data and methods, 317–27. *See also*
    data and methods
  diverse needs, 334
  general interest articles. *See*
    general-interest articles
  speaking to, 282–313. *See also*
    speaking about numbers
  writing for, 314–47. *See also specific*
    formats
  See also specific topics
article. *See* scientific papers and
  reports
associations, 215–20
  bivariate. *See* bivariate associations
  causal. *See* causality
  checklist for writing about, 225
  confounding. *See* confounding
    factors
  correlations, 136, 137t, 216–18
  differences in means across groups,
    217–19
  direction of. *See* direction of
    association
  exceptions in, 221f, 222–23, 223f
  explanations for, 38–40
  interactions and, 222
  magnitude of. *See* magnitude (size)
    of an association
  negative (inverse), 30
  noncausal, 47
  positive (direct), 30
  purpose of describing, 215–16
  spurious, 37, 39

statistical significance. *See* statistical significance
substantive significance, 56–59
types of, 216–20
See also comparisons; direction of
  association; magnitude (size) of
  an association; patterns; three-
  way associations; *and specific*
  types and topics
attributable risk calculations, 57
  causality and, 116
  relative risk and, 117t
  substantive significance and, 115–18
attrition (loss to follow-up), 233–34,
  367n1 (chap. 10)
audiences. *See specific types, formats,*
  and topics
averages. *See* central tendency; mean;
  median; mode
axis labels, 151–52, 158
axis scales, 182–83, 183f, 184, 185f,
  186f, 190

bar charts, 157–63
  to accompany a comparison, 157,
    158f, 217, 221–24, 221f, 223f,
    343, 343f
  clustered. *See* clustered bar charts
  color for, 178, 180
  with consistent and inconsistent
    y-scales, 184, 186f
  criteria for appropriate choice, 6,
    95, 155–63
  error bars and, 175t
  histograms, 156–57, 157f, 159,
    214–15
  horizontal reference lines on,
    171–72
  order of items in, 177, 178f
  shading schemes in, 180, 187, 306,
    308–9

simple, 157–58, 157f
stacked, 156, 160–63, 162f
using line charts when bar charts
are appropriate, 188, 189
See also clustered bar charts;
histograms
baseline, of longitudinal study, 234
baseline of period of observation, for
event history analysis, 119
bell curve (normal distribution), 80f,
114, 163
benchmark calculations, 7, 28
bias, 39–40, 43, 44. *See also*
representativeness
bivariate associations
charts for, 150, 157–68, 174–76t,
189, 291f
interpreting, 319
scientific papers, 248
speaking about, 290
statistical testing of, 220, 261t, 319,
340f
tables for, 125, 134t, 135–37, 135t,
137t, 141
types of, 215, 216, 217, 225
See also correlation; cross-
tabulations
borders, table and cell, 145, 148
box-and-whisker plots, 79, 82, 156,
165–66
briefs, issue or policy
audiences for, 334
charts in, 337–38
color in, 328
contents of, 334
data sources, 318
format, 150, 208, 314, 317, 327,
337–38, 338f
glossary, 336–37
layout and, 335
length of, 337–38

objectives of, 2, 333–38, 344, 345
organization around questions,
335–36
sidebars for, 336, 337, 346
titles, 334–35
W's (who, what, when, where) and,
336
British system of measurement, 16,
69, 199
bulleted lists, 5, 21
in brief presentations, 20
in chartbooks, 327, 332, 339
citations and, 296
in executive summary, 8, 341, 346
indenting and, 295, 297
in policy brief, 337
on posters, 330, 331f, 360t, 364t
in reports, 340
on slides, 285, 286f, 294–95, 301

calculated variables, 216, 238, 240
calculations, explained with analogies, 194–96
case-control studies, 232
categorical variables, 210–13
checklist for writing about distributions and associations, 225
composition of, 210–11
continuous variables and, 82
creating sensible categories, 63–65,
210–11
in data section, 239
defined, 63–64
direction of association between,
157, 159, 161, 174t, 176t, 216,
218
explained in data section, 238, 239,
240
mean and, 210–11, 365n2 (chap. 4)
measure of central tendency of,
365n2 (chap. 4)

digits and decimal places in, 90–91t, 184

error bars on, 166–67, 175t

excessive labeling, 177

explaining "live," 305–10

footnote symbols, 188, 200

in GEE method, 349–50

in general-interest articles, 208, 222, 328, 341, 343f, 344, 346

high/low/close charts, 6, 165–67, 166f, 175t

horizontal reference lines on, 171–72

how many numbers to use, 19–20

for illustrating univariate distributions, 153–57

with inconsistent design of panels, 185

in introductions results and conclusions, 256, 280

legend in, 152–53, 190

linear scale in, 182–83, 183f

logarithmic scale in, 182–83, 183f

maps of numeric data, 168–69, 169f, 180

mixing tools, 20–21

note symbols in, 188

number and types of variables and selection of, 173, 174–76t, 190

number of series, 181

organizing criteria, 177, 178f, 307–8, 349–50

paragraphs for describing complex, 256

point estimates in, 167f

portrait versus landscape layout for, 182

on posters, 329, 330, 360t, 363t

precise values in, 20

for presenting relationships among variables, 157–68

principles for design of, 151–53

reference lines on, 171–72

reference points on, 171

reference regions on, 172–73, 172f

in results section, 260

for scientific papers and reports, 324, 338–40, 340f

self-contained, 151, 190

of sensitivity analysis, 150, 200, 201t, 206

shading schemes in, 180

simplified from research paper, 360t

sizing of, 184, 187

on slides, 287f, 288f, 289f, 291f, 298–301. *See also* slides

statistical significance in, 53, 332. *See also* statistical significance

summarizing patterns, 32–36

three-dimensional effects in, 180

three-way, 355f, 357, 357f

title of, 151–52

topic of, 151, 190

types of, 153–69

units of measurement in, 152, 167, 190

"Vanna White" approach to explaining, 282, 306–7, 310, 312, 333

vertical reference lines on, 172

in writer's toolkit, 6

zero on axis scales, 187, 190

See also bar charts; high/low/close charts; line charts; maps of numeric data; pie charts; scatter charts

checklist

applied audiences, 346–47

basic principles, 36–37, 94

causality, 59–60

charts, 190

against a standard, 84–85, 98
types of, 100–120
wording for, 100, 209
See also analogies; difference
(change); rank; ratios; reference
values
composition, 74f
categorical variables, 210–11
charts to illustrate, 155–57, 174t
histograms for, 156
pie charts and, 6, 150, 154f, 155,
161, 301
stacked bar charts, 156, 161, 163
univariate, 125, 132, 133t, 210, 225.
See also univariate distributions
wording to describe, 210–15
See also univariate distributions
compounding of interest rates, 112
computer software. *See* software
conceptually plausible numeric values
and context, 76–77
effect on choice of example, 76
importance of topic, 76–77
and units of measurement, 76–77
conclusions section. *See* discussion
and conclusions section
concrete examples, 192
concurrent validity, 242
conference bands, 167
conference presentations. *See* speaking about numbers
confidence intervals (CI), 50, 166–67,
175t, 365n3 (chap. 3)
analogy for, 191, 206, 323–24
calculating, 50, 323, 365n3 (chap. 3)
in charts, 50, 166, 167, 167f
high/low/close charts of, 166
interpreting, 323
point estimates and, 166, 167f
standard, 367n3
statistical significance and, 50

writing about, 50, 54, 59, 309, 354
confidence limits, 50
confounding factors, 38
controlling for, 42, 43, 47, 59, 249
definition of, 39
discussed in concluding section, 58,
249, 271
discussed in results section, 116,
290, 290f
consistency checks, 68
for exhaustive categories, 66
for percentages, 66
construct validity, 242
Consumer Price Index (CPI), 84, 98
context
applied audiences, 318
and choice of example, 9, 29, 36, 77,
198, 270
for choosing contrasts, 118–19
conceptually plausible values,
76–77
of measurement, 76
and range of values, 191, 323
of a reported value, 92
in title of table, 122
See also W's (who, what, when,
where)
contingent questions, 236. *See also*
filter questions; skip pattern
continuous variables, 213–15
checklist for writing about distributions and associations, 225
classifying into categories, 62
combined with categorical variables, 82
cutoffs and, 215
defined, 62
interval variables, 62, 103, 105, 118
non-linear patterns, 308, 308f
ratio variables, 62, 106, 107, 118
types of, 62

digit preference and heaping, 205–6
digits
  in charts, 90–91t, 182
  fitting number of, 89–93, 94
  guidelines on number of, 90–91t
  methods for reducing number of,
    92–93
  significant, 87–88, 365n6
  in tables, 90–91t, 143, 149
  See also decimal places
dimensions
  analogies for, 17
  identifying dimensions of a com-
    parison, 350–51
  of units, 69–70
direct (positive) association, 30
direction of association
  in characterizing patterns, 352–54
  exceptions in, 35, 221f, 222, 354–55
  specifying, 30–31, 215, 225, 319
discussion and conclusions section,
    270–76
  abstracts and, 277
  causality in, 271
  checklist for, 280–81, 312–13
  citations in, 271, 272
  as closing arguments, 255
  data and methods and, 227–28,
    248–51
  directions for future research, 250,
    253, 272, 327, 330, 331f, 364t
  limitations discussed in, 281
  numeric information in, 270–71
  organization of, 255
  scientific audiences and, 53
  scientific papers and reports,
    270–75
  slides in, 291, 292f, 294
  statistical significance in, 55–56,
    270–71. *See also* statistical
    significance

strengths discussed in, 281
substantive significance and, 57,
  270–71, 280. *See also* substantive
  significance
distributions, 215–20
  charts for illustrating, 153–57, 174t,
    319
  checklist for writing about, 224
  examining, 76–83
  normal distribution (bell curve),
    80f, 114, 163
  polarized bimodal, 81f
  skewed, 81f
  tables for presenting, 132, 133t,
    135, 135t
  uniform, 80f
  univariate. *See* univariate
    distributions
  wording to describe, 210–15
  See also central tendency; *and*
    specific types and topics
doctoral dissertation, 27, 229, 259
documentation
  accompanying data sources, 232
  book and, 229
  citation and, 131, 230, 237, 296
  dissertation and, 229
  for public release of data set, 229
  sampling weights and, 247, 248
Dow Jones Industrial Average, 119

effect size, 52, 271
effects modifications (interactions),
  223–24
elections, 105, 119
electronic cigarettes (e-cigarettes),
  45–47
empirical criteria
  and charts, 150–51, 158, 190, 349
  comparisons and, 205
  and tables, 349

epidemiological (Hispanic) paradox, 273

equations, use of
  and nonscientific audiences, 25, 247–48, 323
  for scientific audiences, 27, 317, 363t

error bars, 166–67, 175t, 367n2 (chap. 7)

examples, 191–206
  analogies and. *See* analogies
  atypical values in, 202
  audience and, 196–200, 206
  checklist for, 206
  comparability of, 19, 198–99
  for comparing to previous statistics, 192
  concrete, 192
  criteria for choosing, 16–19, 196–201
  decimal system biases in, 204–6
  digit preference and heaping in, 205–6
  for establishing importance of topic, 191–92
  familiarity of, 196–97, 206
  GEE and. *See* "generalization, example, exceptions" (GEE) approach
  ignoring distribution of variables, 202–3
  for illustrating repercussions of analytic results, 192
  objectives of, 17–18
  out-of-range values in, 203
  pitfalls in choosing, 202–6
  plausibility of, 18–19, 197
  reasons for using, 191–92
  relevance of, 19, 197–98
  representative, 351
  simplicity of, 18, 196–97
  single-unit contrasts in, 204

for summarizing patterns, 33–35
  ten-unit contrasts in, 204–5
  units of measurement in, 199
  unrealistic contrasts in, 202–3
  vocabulary of, 197, 206
  See also "generalization, example, exceptions" (GEE) approach

Excel. *See* spreadsheets

exceptions
  in direction, 221f, 222, 354–55
  identifying, 354–56
  interactions and, 221f, 223–24, 223f, 320
  in magnitude, 222–23, 223f, 355
  in statistical significance, 223, 355–56
  for summarizing patterns, 33, 34–35, 35–36
  wording to introduce, 224
  See also "generalization, example, exceptions" (GEE) approach

exclusion of outliers, 241–43

executive summaries, 55–56, 340–41, 342, 346

exhaustive categories, 64

expected change, 76

experimental design, 41–42, 231
  and confounding, 38–39, 42, 43, 58, 116, 249, 271, 324
  control group, 32, 58, 231
  quasi-experimental studies, 42
  treatment group, 32, 58, 231, 249

expository writing, 1–2, 10, 255–56, 280. *See also* prose

extreme values, 76

face validity, 240–41

familiarity of examples and analogies, 196–97, 206

filter questions, 235–36. *See also* skip pattern

final analytic sample. *See* sample
first quartile value (Q1). *See* quartiles
five number summary, 81
follow-up, loss to, 233–34
footnotes. *See* notes
forecasting models, 315, 316, 321
fractions, 70–75
  versus "the whole," 71–72
  See also percentages; ratios
frequency distributions. *See specific
    distributions*
F-statistics, 217
future research, directions for,
    250–51

GEE. *See* "generalization, example,
    exceptions" (GEE) approach
general-interest articles, 52, 192, 208,
    210, 346
  numeric facts, 341, 344–45
  structure of, 223–24, 255, 327, 328,
    341
  technical information on data and
    methods in, 228, 229
  Twin Towers example, 341, 344–45
"generalization, example, exceptions"
    (GEE) approach, 33–36, 220–24
  applied audiences, 321
  characterizing the pattern, 351–54
  choosing representative example,
    351
  displaying data in table or chart,
    349–50
  in explaining graphic images "live,"
    308–9
  identifying dimension of compari-
    son, 350–60
  identifying exceptions, 354–56
  implementing, 349–58
  interactions and, 223
  literature review and, 259

phrasing for, 223–25
  in results section, 260
  slide for, 292f
  speaker's notes and, 284–85
  three-way associations and, 268
  uses of, 33
  writing the description, 356–57
  See also examples; exceptions
generalizations
  phrasing for, 224
  representative examples for, 351
  and statistical significance, 223
  for summarizing patterns, 33–34,
    357–58
general principles, 9, 13–19
geographic attributes, maps for con-
    veying, 168
glossary, 336–38, 347
grant proposals, 315
  data and methods information in,
    229, 252
  general structure of, 222, 255,
    258–59
  introductions for, 225
  literature reviews in, 258
  See also scientific papers and
    reports
grant writers, 314, 315
graphic images
  explaining "live," 305–10
  on slides, 293, 297–98
  See also charts; diagrams; tables
graphing programs, 181, 183
graphs, 170, 173
grayscale, 328
Greek letters, 23
growth rates, 112, 366n4

handouts, 180, 302–3
  to accompany poster, 328, 329, 332,
    333, 363t

to accompany speech, 180, 285,
  302, 303
briefs and, 328
chartbooks and, 328
color and, 302, 328
grayscale, 328
from slides, 180, 285, 302, 303, 362t
software and, 332
heaping, 205–6, 205f
high/low/close charts, 6, 165–67,
  166f
Hispanic paradox, 273
histograms, 83, 156–57, 157f, 159,
  214–15, 214f
horizontal reference lines on charts,
  171–72
how, in data section, 230
human subjects, 131, 230
hypotheses
  causal statements in, 38, 41, 290f,
    297
  direction of association, 30
  literature review and, 280
  speaking about, 330, 331f, 363t
  testing, 316
  See also research question

imputation, 234, 243
indenting
  on slides, 295–97
  in tables, 127–28, 149
independent variables, 140
  categorical, 340f
  causation and, 38, 216
  continuous, 172, 176t
  correlated, 261t, 354
  See also dependent variable
indexes
  appendices and, 127
  construction, 67, 68
  in data and methods, 82, 94

definition, 67
distribution of, 82
scales and, 67, 365n3 (chap. 4)
  See also scales
infant mortality, 204–5
inferential statistics. *See* statistical
  tests
institutional review board, 230–31
integers, digits and decimal places
  in, 90
interactions (effects modifications),
  223–24, 320
interest rates, 112
interior cells of tables, 131, 147
interpreting raw numbers, 28–30
interquartile range, 82, 165, 166f,
  213–14
interval variables, 62, 106, 118
interventions, causality as basis for,
  42–43
introductions, 256–58
  checklist for, 280
  as opening arguments, 255
  to scientific papers and reports,
    257–58
  slide for, 296f
inverse (negative) association, 30
issue briefs. *See* briefs, issue or policy
item nonresponse. *See* nonresponse
items, questionnaire, 140–41, 237,
  240, 352

jargon, 9
  applied audiences and, 317
  briefs and, 337
  in introductions, results and con-
    clusion, 280
  in names for numbers, 21–22
  scanning work for, 24
  in speaking about numbers, 311
  when to avoid, 24–27

as descriptive statistic, 48
differences across groups, 217–19, 218t, 220
geometric vs. arithmetic, 365n5
as not always representing distribution well, 79–81
standard deviation from, 79–82, 114–15, 165, 213, 225
z-scores and, 114–15
meanings, multiple, 22–23
measurement bias, 40
measurement precision. *See* precision of measurement
measurement systems. *See* systems of measurement; units of measurement
mechanisms, causal. *See* causality
median
defined, 79
as descriptive statistic, 48
in five number summary, 81–82
percentile and, 101
memo, 20, 337
metaphors, 192. *See also* analogies
methods section
checklist for, 248
equations in, 247
example of scientific paper, 248
types of statistical methods in, 246–47, 248
weighted data in, 247
See also data and methods
metric system, 16, 69
Million Man March, 17
minimum value, 78, 82, 213
missing by design
codes for, 65
and filter questions, 235–36
and recall bias, 78, 236, 251
and split-sample topic modules, 236
See also skip pattern

missing values, 233, 235, 243
codes for, 65, 78, 230, 241t, 242, 243, 366n4
"missing value" category, 235, 243
recall bias and, 78
skip pattern and, 78
See also imputation
mixed-format slides, 301
mixing tools, 20–21
mode
categorical variables, 99, 210–12, 225, 365n2 (chap. 4)
defined, 79
as descriptive statististic, 48
module. *See* topic module
monetary values, decimal places in, 89, 91t
multiple-line charts, 152, 164–65
comparisons on, 350–51
different y-scales, 164, 165f
generalizing patterns in, 33f, 350f, 353
number of series, 180
multiple-response variables, 66, 155, 158, 175t
multiples, 108t
multivariate models
and confounding, 271
and writing about associations, 210
mutually exclusive categories
applied audiences and, 321
and chart design, 36
defined, 64, 66
and table design, 127

named cohorts, 14
named periods, 14
negative (inverse) association, 30
newspaper articles, 5, 7, 334

random error, 39, 49
randomized clinical trial, 231
random sampling. *See* sampling
range of values, 78
  conceptually plausible, 76, 77,
    190
  definitional limits, 77, 172, 187
  observed, 206
rank
  advantages and disadvantages of,
    103, 105
  for conveying value's position in
    overall distribution, 81
  in quantitative comparison, 97,
    100, 102t
  size of difference not indicated by,
    103, 105
  units of measurement in, 101–2
  and variable type, 118
  when to use, 119
  words for reporting, 103
  See also percentile
rates, 73–75
  annual, 111–12
  death, 75
  defined, 73
  growth, 111–12
  interest, 112
  velocity, 75
rational numbers, digits and decimal
    places in, 90
ratios
  application to US population data,
    104t
  calculation of, 106, 107–8
  common errors when describing,
    107, 108
  for conveying magnitude of as-
    sociation, 225
  decimal places in, 91t

defined, 71
"flipping over," 366n3 (chap. 5)
odds ratios, 194–96
and percentage change, 119
percentage difference as variant
  of, 110
phrases for describing, 107, 108t
in quantitative comparisons, 97,
  100, 102t, 106–8, 119
ratio variables, 62, 106, 118. *See also*
  continuous variables
reporting value for putting in
  context, 118
wording for, 106–7
See also percentages; proportions
ratio variables, 62, 106, 118
raw numbers
  describing variables in data section,
    238–39
  examples of, 114t
  interpreting, 28–32
  omitting in tables, 142
  why report, 30
  See also variables
reading about numbers, 2
recall bias
  and missing by design, 78
  skip pattern and, 78
reference data sources, reports serv-
    ing as, 340
reference lines, 171–72
  in chartbook, 339
  on slides, 302, 309
reference lines on charts, 171–72
reference points, 171
reference regions, 172–73, 172f
reference values, 98–100, 102t
  comparing continuous variable
    values against, 215
  and data availability, 99

reverse causation, 38–39
risk
    attributable risk calculations, 57,
        115–18, 117t
    quantitative comparisons for
        describing, 119
    relative risk, 115–18, 117t, 258, 265,
        288f, 366n6
rounding, 92–93
    to appropriate number of decimal
        places, 89
    to approximate scale, 88
    of computerized output, 87
    human biases in, 205
row labels, in tables, 125–29
RR. See relative risk

sample
    describing, 243–46
    and exclusion criteria, 243–44, 245,
        246
    final analytic, 233–35, 244
    and representativeness, 244
    research question and, 234–35,
        244
    size. See sample size
    See also missing by design; missing
        values
sample size
    data section and, 252
    exclusion criteria, 243–45, 246
    final analytic, 233–35, 244
    longitudinal study, 235f
    minimum and statistical power,
        244
    minimum for subgroup, 236, 237
    and standard error, 50
    statistical tests and, 50
    versus universe (or population),
        233, 243

weights and, 244, 247
    See also attrition (loss to follow-up)
sampling
    biases in, 40
    borderline statistical significance
        and sample size, 52
    in data section, 231–32, 233
    random error in, 49
    representativeness of, 215–16,
        231–32, 244, 351
    sampling weights, 244, 247, 253
    target population and, 132
    types of, 231–32
    See also study design
scale (order of magnitude)
    analogies for illustrating, 192–93
    changing to reduce number of
        digits, 92–93
    definitions of, 365n3 (chap. 4)
    effect of change in, 70t
    in tables, 143
    transforming variables, 82, 83f
    of units of measurement, 16, 69
scales (composite measures), 83
    appendices and, 127
    categorical items in, 67
    CESD scale, 67
    combining variables and, 82–83,
        83f
    construction, 67–68, 69, 82, 94,
        241t
    in data and methods, 238
    definition, 67
    distribution of, 82
    errors in construction, 184, 186f
    indexes and, 67
    Likert scale, 119
    ranges of values, 67
    standardized scores and, 67
    See also indexes

in conclusions section, 272
cross-sectional, 42, 230, 231
in data section, 231, 253, 277, 287
for demonstrating causality, 44–47
exclusion criteria, 243
keywords for, 276, 278–79
longitudinal, 230, 235f, 297
and missing values, 243, 287, 366n4
prospective, 231
retrospective, 231
split sample module, 236
title and, 276
types of, 4, 231
See also experimental design
study guide, 10
study nonresponse. *See* nonresponse
substantive meaning. *See* conceptually plausible numeric values
substantive significance, 56–59
applied audiences and, 271
attributable risk and, 115–18
causality and, 58–59
checklist for, 59–60
costs/benefits analysis, 326
definition of, 56–57
examples for, 56–57
prevalence and, 56, 115
in results section, 271
statistical significance and, 56–58,
58–59
substantive importance and, 326
summarizing patterns, 32–36. *See also*
"generalization, example, exceptions" (GEE) approach
survey data, nonreponse in, 242
symbols
applied audiences, 4, 24, 36, 194,
317, 323
on charts, 170
defining, 21
Greek, 23, 25, 317

on maps, 169
plotting, 180, 188
on slides, 302–3
for statistical significance, 136, 298,
324, 354
in tables, 145, 147, 149
synonyms, 23–24
systems of measurement
British system, 16, 69, 199
metric system, 16, 69
varying, 16, 69–70
See also units of measurement

tables, 121–49
abbreviations and acronyms in,
126, 131–32
adding dimension to, 128
alignment of components of,
143–44
alphabetical ordering of items in,
140, 141, 149
anatomy of, 122–32
appendix, 129, 144, 149
applied audiences, 324
bivariate, 125, 134t, 135–37, 135t,
137t
borders in, 145, 148
in brief presentations, 20
checklist for, 148–49
column headings in, 129, 131, 149
column spanners in, 131, 138t, 148
column widths for, 147, 148
conceptually related blocks in, 129,
130t
context of, 123, 148
coordinating with writing, 139–45
decimal places in, 89, 90–91t, 93,
143, 149
different measures for same relationship in, 128–29
drafting, 145–48

tables (*continued*)

empirical ordering of items in, 139,
140, 141, 149
explaining "live," 309–10
focused, 122
formatting for different publications, 145
in general-interest articles, 341
in "generalization, example, exceptions" approach, 349–50
how many numbers to use, 19–20
indenting row labels in, 127–28,
149
interior cells of, 131, 147
in introductions, 256, 257, 280
layout of, 145–48
mixing tools, 21
notes to, 131–32, 149
order of importance of items in,
139–41
organizing criteria for, 139–41,
349–50
panels in, 128–29, 130t, 139, 148
paper and pencil for drafting, 146
paragraphs for describing complex,
256
planning principles for, 121
portrait versus landscape layout
for, 144, 148
questionnaire items' ordering in,
140–41
results and conclusions, 260–72,
280
row labels in, 125–29
scale of numbers in, 143
self-contained, 122
shape and size determination for,
146–47
simplified from research paper,
360t
on slides, 288f, 298–301

summarizing patterns, 32–36
technical considerations for,
141–45
theoretical grouping of items in,
139, 141, 149
three-way, 136, 138t, 350–51, 352t
title of, 122–23, 125, 148, 272–76
topic of, 123, 148
"total" rows in, 141
type size for, 144
types of, 132–36
units of measurement in, 123, 125
univariate, 125, 132, 133t, 135
"Vanna White" approach to explaining, 282, 306–7, 310, 313,
333
variable names from statistical
packages in, 126
which numbers to include, 141–43
word processors for creating, 145,
147–48
in writer's toolkit, 5–6
See also cross-tabulations
target population, 132
technical terms
data sources and, 318
defining, 21–24
glossary for, 336–37
notes and, 118
when appropriate, 24–28
when to rely on, 27–28
See also jargon; terminology
temporal ordering. *See* time
ten-unit contrasts, 204–5
terciles, 103. *See also* rank
terminology
ambiguous, 22–23
for applied audiences, 4, 22, 197,
318, 323
for associations, 22, 26
briefs and, 336, 337, 338f

charts and, 189

definitions and, 21, 313, 318, 341

differences across disciplines, 231, 365n3 (chap. 4)

general-interest articles, 210, 341

jargon and, 248. *See also* jargon

mutual exclusivity and, 127

proportions and, 71

tables and, 122

See also specific terms and topics

test scores, standardized. *See* standardized test scores

test statistics, 49, 89, 91t, 92, 324, 330

applied audiences and, 322

decimal places for, 89, 91t, 92, 143

in discussion section, 270

null hypothesis and, 49

p-values and, 52, 53, 303, 324

statistical significance and, 54, 223, 298

tables and, 143

three-way cross-tabulations, 220

See also chi-square statistics; *F*-statistics; statistical tests; *t*-statistics; *z*-scores

text slides, 292f, 294–97

theoretical criteria

for comparisons, 139, 149, 158, 181

for organizing table or chart, 179f, 180, 190, 349–50

thesis statement, 7, 10

third quartile value (Q3), 82

three-dimensional effects, 181

three-way associations

applied audiences, 320

bar chart for, 267f

checklist for writing about, 225

defined, 215

"generalization, example, exceptions" approach to describing, 33–36, 220–24, 268

interactions and, 320. *See also* interactions (effects modifications)

types of, 220

three-way cross-tabulations, 220

three-way tables, 136, 138t, 350–51, 352t

thresholds. *See* cutoffs (thresholds); standards (cutoffs; thresholds)

time

annual rates, 111, 366n4

associations and, 216

benchmarks and, 99

causation and. *See* causality

cohorts and, 14

comparisons and, 18, 92, 101, 198, 350

cross-sectional data, 230

dates. *See* dates

digit preference and, 205

generalization and, 224

longitudinal data, 230, 234, 235f, 297

measuring, 69

numeric examples and, 191

percentage change and, 109–11

plotting and, 164

reliability, 242

temporal ordering, 41–42, 47, 99, 164

timelines, 297

trends, 119, 319, 352–54

W's and, 230, 278

See also specific topics

time allocated for speeches, 283, 286, 293, 305, 306, 310–12

titles

of charts, 151–52

of papers and reports, 278–79

of posters, 332, 361t

of slides, 293–94

range of, 78
typical, 76
See also precise values; reference values; standards
"Vanna White" approach, 282, 306–7, 310, 313. *See also* speaking about numbers
variability in data, 79–82
variables
"alphabet soup" names for, 126
associations. *See* associations
calculated from original data, 238
categorical. *See* categorical variables
causation and, 216. *See also* causality
in charts. *See* charts
checklist for, 94
conceptually plausible values and, 76–77
continuous. *See* continuous variables
creating new, 68, 240, 253
in data section, 238–42, 253
defining, 86, 253, 287, 311
dependent. *See* dependent variable
dichotomous, 142, 157, 174t
explaining in results section, 278, 280
independent. *See* independent variables
level of measurement, 61
levels of, 77
missing value codes, 78
nominal. *See* nominal variables
ordinal. *See* ordinal variables
raw vs. calculated, 238
reporting in data section, 237, 237–38
roles of, 216

in single- versus multiple-response questions, 66, 155, 158
in tables, names of, 126
transforming, 82–83
types of, 61–68. *See also specific types*
W's and, 237
See also specific types and topics
variance, 79, 94, 174t, 217, 246
vertical reference lines on charts, 172
vocabulary
about causality, 43–44
of examples and analogies, 197, 206
See also defining terms; jargon; technical terms; terminology

website, 10
weights, sampling, 247, 253
word problems, 1, 7
word processing programs, 145, 147, 148, 305
writing about numbers
abstract ideas and, 9
causality and significance
causality, 37–48. *See also* causality
checklist for, 59–60
statistical significance, 47–56. *See also* statistical significance
substantive significance, 56–59. *See also* substantive significance
characteristics of
different ways of, 1
as expository writing, 1–2
identifying the role of the numbers, 7
as iterative process, 7–8
legal argument compared with, 255

Made in the USA
Lexington, KY
29 August 2018